# Building
# Your Own
# Airplane

## AN INTRODUCTION

# Building
# Your Own
# Airplane

## AN INTRODUCTION

## DONALD H. WALTER

Iowa State University Press / Ames

THIS BOOK is dedicated to
the wonderful women in my life:

*my wife, June, and my five daughters—*

*Patricia, Diane, Carolyn, Elizabeth, and Fawn.*

*They have put up with my being crazy about airplanes*

*for as long as they can remember,*

*and I love them all.*

---

**Lt. Col. Donald H. Walter** (USAF retd.) has been a fighter pilot, chief of a Minuteman silo crew and an aircraft and helicopter mechanic. He has a master's degree in aerospace engineering and also holds an airframe and powerplant mechanic's license. Currently, Don Walter lives in Riverside, California, where he is an EAA technical counselor and is engaged in building his own Barracuda.

The purpose of this book is to provide information on building an airplane from plans. The user of this information assumes all risk and liability arising from such use. Neither Iowa State University Press nor the author can take responsibility for the actual operation of a homebuilt aircraft or the safety of its occupants.

© 1995 Iowa State University Press, Ames, Iowa 50014
All rights reserved

♾ Printed on acid-free paper in the United States of America

First edition, 1995

Library of Congress Cataloging-in-Publication

Walter, Donald H.
    Building your own airplane: an introduction/Donald H. Walter.—1st ed.
        p.          cm.
    Includes bibliographical references and index.
    ISBN 0-8138-2793-0 (acid-free paper)
    1. Airplanes, Home-built.  I. Title.
TL671.2.W254      1995
629.133'343—dc20                          95-7356

# CONTENTS

# PREFACE

A NUMBER OF YEARS AGO, AS A NEWLY RE- TIRED AIR FORCE fighter pilot, I wondered how I could continue to indulge my avid desire to fly even though Uncle Sam was no longer providing Mach 2 toys for me to play with. I knew that airplanes, even small puddle jumpers, were expensive to buy, expensive to operate, and expensive to maintain. I also knew if I was going to own a flying money pit, I would have to be able to maintain it myself, so I used GI Bill funding to attend the airframe and powerplant (A&P) course at the Northrop Institute of Technology. After completing the A&P course and the necessary Federal Aviation Administration (FAA) exams, I had my aircraft mechanic's license in my pocket. Now, all I needed was the money to buy a plane. In order to earn that money, I went out into the big wide world looking for a flying job. I soon found that there were at least ten pilots for every flying job available. What's more, the companies wanted a twenty-five-year-old pilot with 10,000 hours of diverse flying time (heavy on multiengine jets) and all the ratings in the world. Having instilled in myself and my family the nagging habit of eating three times a day, I was forced to take a job as an A&P mechanic when it became painfully obvious that absolutely no one was going to pay me to fly.

At Northrop, I had heard about a bunch of crazy nuts who were actually out there trying to build their own airplanes. Most of these aviation screwballs were in an organization known as the Experimental Aircraft Association (EAA). Renting a Cessna 150 to get in the air once in a while was proving to be less than satisfactory, so I decided to check out this do-it-yourself bunch. I found there were four EAA chapters in my local area, and I dropped in on one of their meetings. To my considerable surprise, I discovered that the members weren't nut cases after all and that many homebuilt aircraft were far superior in quality, appearance, and performance to comparable factory-built planes. Most important, from my standpoint, the homebuilts were far, far cheaper, and the builders themselves could maintain, modify, and inspect them.

Since aircraft mechanics aren't known for amassing large quantities of wealth, I decided that if I was going to obtain a relatively high-performance aerobatic airplane, I would have to build it myself. I jumped in with both feet, looked over the available models, and ordered a set of plans. I selected a Jeffair Barracuda, a two-seat, all-wood, low-wing plane capable of limited aerobatics (no snap maneuvers) and a 200-MPH top speed. As I progressed in my construction, I gained some knowledge about homebuilt aircraft, made some mistakes, and uncovered some pitfalls. This book pulls together in one place what it has taken me a number of years to learn from numerous and varied sources. I hope it will prevent someone from making the same mistakes I made.

# Building
# Your Own
# Airplane
## AN INTRODUCTION

This Steen Skybolt is a beautiful example of the craftsmanship that goes into a homebuilt aircraft.

**The Montgolfier balloon was prime entertainment in eighteenth-century France.**
Photograph courtesy of the Experimental Aircraft Association.

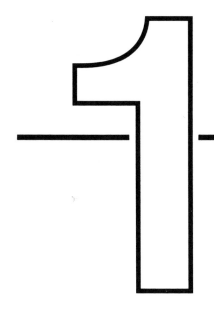

# History of Homebuilt Aircraft

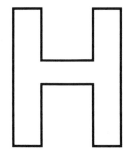

**OMEBUILT AIRCRAFT go back a long way. Icarus's inventive father fashioned wings of feathers and wax, but the impetuous youth ventured too high and the sun melted the wax (the mythic Greeks weren't too well versed in the physics of the atmosphere), with predictable and fatal results. Later, Leonardo da Vinci came up with some interesting designs and models for flight, but it was ballooning in the 1700s and 1800s that really got the homebuilt movement off the ground, literally and figuratively.**

The Montgolfier bothers, Jacques and Joseph, of France are credited with the honor of being the first to send an unlikely trio—a duck, a rooster, and a sheep—into the air under a cloth bag lined with paper and filled with hot air from a pan of burning charcoal. Obviously, homebuilts attracted some high-level attention in those early days, since the king of France, Louis XVI, came out to witness this auspicious event. On October 15, 1783, Jean Francois Pilatre de Rozier, the king's historian, became the first human to go aloft in a homebuilt aircraft (tethered to the ground though it was), where he stayed for four and one-half minutes, reaching an altitude of 84 feet. The following month, Pilatre de Rozier and the Marquis d'Arlandes rose in an untethered balloon to a height of 500 feet on a 25-minute flight over the city of Lyon. The homebuilt movement was off to a flying start.

The Montgolfiers' cloth and paper hot-air balloon had one significant drawback: sparks from the fire could set the bag alight, and this always led to a rapid descent followed by a very sudden stop at ground level. Another Frenchman, physicist J. A. C. Charles, who developed Charles' Law of Gases, was experimenting with hydrogen-filled balloons. A varnished silk was used to contain the hydrogen, and when one such un-manned balloon came down in a field outside Paris, frightened farmers attacked the "evil spirit" and destroyed it. King Louis had to issue a description of balloons

**The brothers Montgolfier get ready for liftoff.** Photograph courtesy of the National Air and Space Museum, Smithsonian Institution.

to his subjects and instruct them not to attack them.

In December 1783, Charles and one of the Robert brothers, who had developed the varnished silk material for the gas bag, ascended in a hydrogen-filled balloon and were carried more than 25 miles from Paris. When they landed, Robert stepped out of the basket to allow Charles to make a solo flight. The balloon rose rapidly to an altitude of 9,000 feet, as recorded on the onboard barometer, and Charles discovered, to his considerable discomfort, that the air is very thin and very cold at such heights.

Ballooning continued to develop at a rapid pace. In 1793, President George Washington watched Francois Blanchard make the first manned balloon flight in the United States. Balloons were used for observation during the Civil War and in numerous wars in Europe. The English Channel was crossed by balloon, and much later, even the Atlantic and the Pacific oceans were crossed by free-flight balloons. On a sad note, Pilatre de Rozier, who made the first flight, died in the explosion of a combination hydrogen

and hot-air balloon in 1785. Balloons had gotten people off the ground and into the sea of air above the earth, but the scientists, the inventors, the tinkerers, and the builders wanted more control over their destinies and their destinations. They weren't content to drift wherever the wind might take them, and so, the race was on to be the first to fly a heavier-than-air machine that could be controlled in flight.

The early homebuilders were an ingenious lot and tried a multitude of methods for propelling themselves into the air. Some tried to copy the flight of birds and designed and built human-powered ornithopters with wings that flapped. Not having hollow bones and fluffy feathers proved to be a major disadvantage, and flap and flail as hard as they might, their attempts still simulated streamlined rocks rather than birds.

Some tried to duplicate the design of Leonardo's helicopters, but there was no suitable engine available. Moreover, the concepts of Coriolis effect, retreating blade stall, coning effect, leading and lagging blades, and airfoil effects were not clearly understood at that time. The helicopter would have to wait for the 1930s and for Henrich Focke of Germany and Igor Sikorsky, a Russian engineer who came to the United States in 1919 to develop the first practical helicopters.

Fixed-wing designers and builders did not have much better luck at first. They were handicapped by the lack of a suitable powerplant and a lack of understanding of the principles of aerodynamics. In England, Sir George Cayley constructed a homebuilt with two circular wings and a steam engine driving two huge propellers. It didn't have a chance of the proverbial snowball in Hades of getting off the ground. Another Englishman, William Henson, put together a model of his steam-powered plane that had a 40-foot wingspan (the full-scale plane was to have a 150-foot span), but his structure was a little too fragile. When dew settled on the silk covering overnight, the cover shrank and warped the wings completely out of shape. A few years later, Henson's partner, John Stringfellow, built a model with a 10-foot wingspan and a steam engine that carried the model 120 feet through the air.

Other homebuilts were coming together in other countries as well. In fact, a French electrical engineer, Clement Ader, claimed to have flown his Eole design and later his Avion, but both flights ended in crashes that destroyed the aircraft. Ader survived the crashes and lived to see two American bicycle mechanics perform more successful flights.

In Germany, Otto Lilienthal was doing some very impressive work with homebuilt gliders. Lilienthal used weight shift control exactly the way some hang gliders do today. His glider flights increased in distance and duration as his experience increased. He had developed a 2.5 HP carbonic acid-burning engine for one of his gliders, but he was killed on a glider flight in gusty wind conditions before he was able to try his powered plane. His family and associates were so distraught by his death that they stored the powered plane in a shed and it was never flown.

Work was going forward in the United States under the guiding hands of Samuel P. Langley, Charles Manly, and Octave Chanute. Chanute, a respected bridge and railroad engineer, studied the problems associated with manned flight and wrote articles that he published as a book, *Progress in Flying Machines*. He constructed a number of gliders with as many as 12 wings. Chanute concentrated on building stability into his designs and didn't develop controls for his gliders or search for an engine to provide powered flight. His pilot, Augustus Herring, tired of flying gliders and went his own way. Herring built a biplane with a compressed-air engine and claimed to have flown it successfully for a few seconds, but this was never verified.

Meanwhile, Langley and his assistant, Charles Manly, were successfully launching models powered by small steam engines off the deck of a houseboat on the Potomac River. His experiments attracted the interest of President McKinley, and government funds were provided for a

manned aircraft. A suitable engine was still not available, so Manly, the consummate assistant, designed and built an internal combustion engine with an astonishing 52 HP output. On the first attempt at flight with Manly at the controls, the plane hit part of the launching catapult and fell into the river in front of a large contingent of government observers and representatives of the press. A second attempt resulted in structural failure during the catapult launch, and again, the homebuilt went into the water. When Manly barely escaped drowning during this attempt, he decided to take up another line of work. The ensuing public ridicule drove Langley back to his work at the Smithsonian.

Two other Americans, bicycle mechanics in Dayton, Ohio, were trying their hands at manned flight. Orville and Wilbur Wright were to become famous for their homebuilt

**Designer Samuel Langley (*right*) and his pilot, Charles Manly. Poor Charlie kept getting dunked in the Potomac River.** Photograph courtesy of the Experimental Aircraft Association.

**Otto Lilienthal takes a flight in one of his early gliders, which looks a lot like some of the hang gliders of today.** Photograph courtesy of the Experimental Aircraft Association.

Wright Flyer, but they had a number of obstacles to overcome first. Just as their contemporaries were doing, they experimented with gliders at first, but they went a few steps further. They built the first wind tunnel, in which they tested their designs and ideas. They developed wing warping as a means of directional control, and as Manly had done, they designed and built their own internal combustion engine. The brothers also developed a movable vertical control that allowed them to make coordinated turns in the air, and they designed and carved their own propellers. Numerous times, their gliding flights ended in the loss of control and crashes that could have been fatal, but they survived with only bumps and bruises. After years of problem solving, trying different approaches, some successes and many failures, Orville Wright flew 120 feet on December 17, 1903, on the cold, windswept sand dunes of Kill Devil Hill, Kitty Hawk, North Carolina. The homebuilt movement (though it wasn't known as that at the time) had passed a major milestone, with many more to follow in rapid succession.

Numerous homebuilders followed in the footsteps of the brothers Wright in countries around the world. In France, Alberto Santos-Dumont, a Brazilian living in Paris,

flew his own design 656 feet. Hubert Latham attempted to fly across the English Channel in his Antoinette monoplane, but he took a swim partway across because of engine failure. Luckily, the Antoinette floated quite well. Louis Blériot of France successfully crossed the channel in his Blériot monoplane in 1909, winning the prize offered by the London *Daily Mail*. In America, Glenn Curtis and Lincoln Beachey were flying their own designs. In Russia, Igor Sikorsky built and flew a four-engine monster called Le Grand. Flights in homebuilts were also taking place in Germany, Sweden, Italy, Austria, Portugal, Romania, and Turkey.

When World War I rolled across Europe, homebuilt aircraft took a back seat to professionally manufactured aircraft, as the nations of the world realized the potential of the flying machines and started turning out warplanes. Fighters, bombers, and trainers rolled off the assembly lines, and driven by man's inhumanity to man, the performance of aircraft increased dramatically.

After World War I, homebuilt designs burst on the scene again. They soon equaled and even surpassed the best the military had to offer. Designer and builder Ed Heath

**The famous first powered flight. The Wright brothers do it right in a homebuilt aircraft.** Photograph courtesy of the Experimental Aircraft Association.

**Louis Blériot in France after the first flight across the English Channel in his homebuilt bird.** Photograph courtesy of the Experimental Aircraft Association.

built and flew his Heath Parasol and his Baby Bullet. The Heath Parasol is still a popular homebuilt today. Bernie Pietenpol built another parasol design, the Pietenpol Air Camper, which is still being built by amateur builders around the world. Vernon Payne's Knight Twister was years ahead of the commercial aircraft being produced at the time. Homebuilts like the Gee-Bee, the Laird Solution and Super Solution, and the Wedell-Williams Racer were routinely winning the Bendix Race and the Cleveland National Air Races. The military and the commercial manufactur-ers could not match their speed, reliability, and endurance.

In 1938, the Civil Aeronautics Administration (CAA), the forerunner of the Federal Aviation Administration (FAA), completely revised its regulations but failed to address the area of amateur-built aircraft. Before this gross oversight could be rectified, World War II was upon us, and all significant effort was directed toward building military fighters, bombers, transports, trainers, and gliders.

At the conclusion of World War II, a hue and cry was raised across the country to reinstate regulations that would

**World War I put a damper on homebuilt aircraft for a while so that factories could turn out war machines such as these Spads.** Photograph courtesy of the National Air and Space Museum, Smithsonian Institution.

This is a replica of the original Gee-Bee. Is that a fast-looking machine or what? Photograph by Tim Koepnick, courtesy of the Experimental Aircraft Association.

allow reestablishment of the homebuilt movement. George Bogardus flew his Little Gee-Bee homebuilt from Oregon to Washington, D.C., to publicize the problem, and the CAA wrote amateur-built aircraft into its experimental licensing category. Thus, homebuilts came to be known officially as experimental aircraft, even though they are not experimental per se, as are the X-1 or the X-15.

In the early 1950s, a small event took place in Milwaukee, Wisconsin, that would grow like Topsy. An ex-P-51 jock, Paul Poberezny, got a few pilots and designers together in the basement of his home and started talking flying, designing, and building. As interest in their group began to spread, they formed the Experimental Aircraft Association (EAA) in 1953 and began publishing a small newsletter. As the organization continued to grow, they set up headquarters at the airport at Hales Corners, Wisconsin, and started an annual fly-in for homebuilts. It wasn't long before the EAA expanded to include kit-built aircraft, antique aircraft, warbirds, classic aircraft, aerobatic aircraft, helicopters, gyrocopters, autogyros, and ultralights. The organization outgrew the facilities at Hales Corners and moved to Oshkosh, Wisconsin. The weeklong EAA Fly-In Convention now makes Oshkosh the busiest airport in the world during those seven days. The staggering statistics for 1994 for the event document more than 12,000 aircraft of

all shapes and sizes and more than 850,000 people there to look them over.

The Oshkosh facilities include a world-class aviation museum showing all facets of aviation and Pioneer Airport,

**George Bogardus, who put homebuilt aircraft back on the FAA books.** Photograph by Owen Billman, courtesy of the Experimental Aircraft Association.

**A wing-walker performs during the daily air show at the annual EAA fly-in at Oshkosh, Wisconsin. It's a really big show.** Photograph courtesy of the Experimental Aircraft Association.

which represents aviation as it was in the 1920s and 1930s. EAA chapters have sprung up around the world while membership has exploded to more than 140,000 in 1994 and still growing. EAA is the mainspring driving the homebuilt movement, but there are also many efforts and activities going on outside the EAA. There are nearly 500 homebuilt designs to chose from and a multitude of companies providing parts, plans, kits, hardware, and supplies for the person who wants to join in the fun. So if the bug has bitten and you want to put together a flying machine with your own two hot little hands, how do you pick one from the massive number available? We will take a look at the selection process next.

**A Barracuda, a two-place, all-wood homebuilt that looks as if it is anxious to get back in the air.**

# 2 Selecting Your Plane

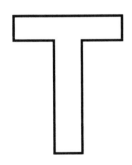

**HERE ARE MANY FACTORS** to consider when you are selecting the plane you want to build. If you are a first-time builder, remember KIS: Keep It Simple! Don't take on a project that is too hard for you to handle or you will greatly reduce your chances of completing it.

You have nearly 500 homebuilts to chose from, but by making a few judicious decisions up front, you can reduce this number to a manageable few and make your final decision from a shorter list.

First and foremost, what material do you want to work with—welded steel tube, wood, aluminum, or composite material (fiberglass and foam)? The second major decision is how many people do you want your winged chariot to carry? A few other factors you will want to consider include these: open cockpit or closed; biwing or single wing; high performance or low; aerobatic or not; land, water, or amphibious; single engine or multiengine; tractor (pulling) engine or pusher; aircraft engine or converted auto engine; and possibly the most important, how much money can you afford to spend?

I'm sure you have heard it said that life is a series of compromises. Well, you will have to compromise on some things when you are deciding on the plane you want to build. Decide what is important to you and what you can live without. For example, I wanted a high-performance plane that was capable of doing some aerobatics. Being an ex-fighter jock, I wanted to get "upside down and backwards" once in a while. In order to achieve this objective, I had to give up low cost, low power, and short building time. You may want to make a priority list. Start out with the characteristics you feel you must have. Next list the things you would like to have. Then note the items that would be nice to have but you could get along without. Finally list the things that don't matter to you one way or the other.

Here is the priority list that led to the selection of the Barracuda for my project:

*Must have:*
    Medium cost spread over construction period
    Two seats
    Aerobatic capability
    High performance
    Closed cockpit
    Good building instructions

A fairly simple all-wood project, a MINI-MAX, under construction. Notice the wings stored overhead.

*Would like:*
   Wood construction
   Low wing
   Control stick
   Single engine
   Land operation
   Sufficient elbowroom in cockpit

*Nice to have:*
   Retractable gear

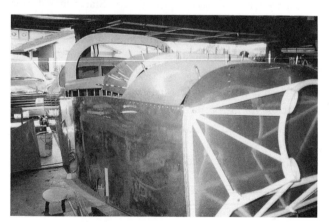

An example of all-metal construction, a Van's RV-6. Clecos hold the sheet metal in place before rivets are bucked.

Simple construction
Short building time
Auto engine conversion capability

*Doesn't matter:*
   Pusher or puller engine
   Conventional or tricycle gear

Let's look at some of the options. As I mentioned above, there are four primary types of construction: welded steel tube, wood, aluminum, and composite. Of course, no plane is all wood or all foam and fiberglass. For instance, a so-called all-wood plane has many steel and aluminum brackets and fixtures. The fuel tanks may be aluminum or fiberglass. The engine mount and landing gear will probably be welded steel tube. In addition, you will have Plexiglas in the canopy and windshield, rubber in seals and tires, and fabric in the seats and upholstery. In every case, your homebuilt will be a mixture of materials, but the primary material is important to you because this is the material you will be working with month after month as construction proceeds.

When deciding on the type of construction you want to use, consider what you are experienced with and what you like to work with, but don't let this limit your decision completely. Don't let the idea of working with steel tube or alu-

**A Breezy clearly shows its welded steel tube construction.**

**Two Long-EZs rest on their noses (the nosewheel cranks down after the pilot and passenger climb in) and show off their smooth-as-glass (fiberglass) composite construction.**

minum sheet metal or fiberglass intimidate you. Even if you have no experience with these materials, you can learn and you can get help. Your local EAA chapter members will give you a hand. Your local community college or trade school offers courses in welding, sheet metal, and woodworking. The EAA and aviation shops will be glad to sell you how-to books on all facets of homebuilt construction. It will give you a great deal of personal satisfaction when you hold a well-made part in your hand whether it is welded, riveted, cut from wood, or laid up with fiberglass. You may create some scrap parts as you are learning, but believe me, you can learn.

Is one type of construction better than another? That depends on whom you ask. Each type has advantages and disadvantages. Moisture can cause wood to rot and steel to rust. Oxidation can eat into improperly protected aluminum. Ultraviolet rays from the sun can cause fiberglass to deteriorate. Each material has builders who swear by it and others who swear at it. If proper construction procedures are followed and the completed plane is maintained and inspected properly, all four materials will result in a homebuilt that will give you many years of flying pleasure.

Available tools may affect your decision on the material you wish to use. Many people have woodworking tools, either hand tools or power tools, but few have welding rigs or sheet metal tools. Table 2.1 shows a list of basic tools needed to work with the four types of material. Costs will vary greatly depending on the quality of the tools you pur-

**TABLE 2.1. Special Tools Required for Various Construction Methods**

| Woodworking Tools | Steel Tube Construction Tools | Aluminum Construction Tools | Composite Construction Tools |
|---|---|---|---|
| ⅜-inch electric drill | welding rig | rivet gun | scissors |
| ½-inch drill press | hacksaw | rivet squeezer | squeegee |
| hacksaw | clamps | rivet sets | mask |
| table saw (or radial arm) | set of files | bucking bars | filters for mask |
| fine-tooth model saw | 3-inch vice | blind rivet tool | hot wire foam cutter |
| wood rasp | wire brush | rivet cutter | |
| magnetic tack hammer | | set of files | |
| electric stapler | | 3-inch vice | |
| clamps | | many Clecos | |
| wood vise | | Cleco pliers | |
| hole-cutting bits (for holes | | ⅜-inch electric drill | |
|   over ½ inch) | | counter sink | |
| drill bit set | | drill bit set | |
| | | ½-inch drill press | |
| | | chip chaser | |
| | | hole-finder set | |
| | | deburring tool | |
| | | tin snips | |
| | | angle drill with bits | |

chase. You don't need gold-plated tools, but be sure you get good-quality tools. Poor tools will make your job more difficult, will give poorer results, and will not hold up very long.

One additional type of construction is used strictly for ultralights, planes weighing not more than 254 pounds empty weight and not requiring FAA certification or a license for the pilot: this is aluminum tube construction. If you decide to build an ultralight, you will find this type of construction simple and relatively inexpensive.

The number of seats your homebuilt will carry is an important consideration and should be given some serious thought. Obviously, the more seats you require, the larger the airframe, the larger the engine, the longer the building time, and the more the overall costs. If you just want to get into the air by yourself to enjoy the spectacular pleasures of flight, then a single-place plane is the one for you. If, on the other hand, you want to share these pleasures with a friend, a spouse, your kids, or anyone at all; you will need two or more seats. Three- and four-place homebuilts are rare, but they do exist. Here you have to dust off your crystal ball and look into the future. Singles may have a spouse and a bunch of rug-rats in their future. Rug-rats may grow up to be budding pilots wanting to get some stick time. A two-place plane is a safe way to go. You can't take up a whole group, but you can share your experiences with one other person at a time. Or, you can go up by yourself to commune with the sky and the clouds when you want to be alone. As I said before, give this one a lot of thought before you start ordering plans, kits, or materials.

Do you dream of the "good old days" when barn-stormers crisscrossed the country in their surplus Jennies? Do you want to hear the wind whistling through the wing wires, or do you prefer the comfort of a heated, enclosed cockpit? In other words, do you want an open-cockpit biplane, a canopy-covered biplane, a low-wing, high-wing, or maybe a mid-wing? You may have learned to fly in a high-wing trainer and may wish to continue to fly this type. Maybe you are an old fighter jock (or a wannabe) and you have to have a low-wing plane with a stick instead of a control wheel. Maybe it doesn't make any difference to you where the wings are as long as they don't fall off. But for most builders, it is another important consideration. Remember, a biwing will double the number of ribs and spars you must build and assemble into wings to be covered and installed, but this is not an insurmountable obstacle if this is your heart's desire.

Next, let's look at the feet you want to have under your little beauty. Do you want a taildragger, also known as conventional landing gear, or a nosedragger, sometimes called tricycle landing gear? Possibly, you live in the land of lakes or anyplace with a lot of water around. Then you may want to consider a floatplane or, for more flexibility, an amphibian. There are a few things to consider here. If you have only flown planes with a training wheel (nosewheel), a tail-dragger will require a little flying time with an instructor to get the hang of it. You can't land a floatplane or an amphibian in any damp spot you find since the FAA has to approve water landing areas. Airports for floatplanes are limited, and this factor will restrict to some degree where you can go on a cross-country flight.

This Challenger ultralight rests on its tail until the pilot gets into the cockpit.

This Stolp SA-100 requires a lot of rib building, but a constant chord wing simplifies things since most of the ribs are identical in size and shape.

**The Osprey offers the versatility of landing on land or water.**

Now, how fast do you want to go? Do you prefer low and slow so that you can enjoy watching the countryside pass beneath your wings, or are you a speed freak? The SX-300 will do just what its name implies, get you from point A to point B at 300 MPH. An ultralight will putt along at about 40 MPH and you can watch the cars pass you on the road below. There is a supersonic homebuilt (the BD-10) in the works, but the quarter-million-dollar price tag puts this speed machine out of the reach of most aspiring builders. Between 40 and 300 MPH, there should be something to fit your tastes.

There is another aspect of performance besides raw speed. Are straight and level flight and 30-degree-bank

**The Prowler is a real speed demon, capable of 300 MPH.**

turns your cup of tea, or do you like to turn every way but loose? Let's take a look at g-loading and aerobatic capabilities. In level, unaccelerated flight, your plane and your body encounter one g, or the normal force of gravity. In a turn, this may go up to two g's, or twice the force of gravity. Now the weight of your plane has doubled. Make a hard landing with eight g's and your wings may wind up on the ground with the gear sticking up through them. If you are planning on doing some aerobatics, the plane you choose must be designed and built to carry the g loads for the maneuvers you wish to perform. According to Appendix A of Part 23 of the Federal Aviation Regulations (FARs), a normal category plane (no aerobatics) requires a design that will sustain +3.8 g's. A utility category plane designed to perform limited aerobatics (no snap maneuvers such as snap rolls) requires +4.4 g's. If you want to go all out with every aerobatic flip and turn in the book, your bird will need at least +6 g's and an inverted fuel and oil system for maneuvers involving sustained negative g's. Whatever your choice, there is a homebuilt design that will stay together while you wring it out. Be sure to give your passengers a strong plastic bag before you turn them upside down. It will make cleanup much easier.

If you have made the decisions discussed above, the size of the powerplant will pretty well be limited to a certain range of horsepower. You may need a pusher or a tractor type, or one of each. Your engine (or engines) may be large or small, new or used or rebuilt, low time or high time, but there is one more decision you can make. You may select an engine built specifically for aircraft use or an engine built for some other purpose and modified for aircraft use. The latter type includes automobile engines, auxiliary power unit (APU) engines, and snowmobile engines. For many years, Volkswagen engines have been used on smaller homebuilts, as have APU and snowmobile engines. In recent years, larger auto engines—such as Ford V-6s and Buick V-8s, Corvair, Mazda, and Subaru engines—have been converted to aircraft use with reduction drives to reduce propeller RPM to a number that will keep the prop tip from going supersonic. Usually, considerable money can be saved by using a nonaviation engine-type if there is one compatible with the homebuilt design you select.

Here's something else to think about. Where are you going to build this marvelous flying machine? Homebuilts have been constructed in bedrooms, dining rooms, living rooms, basements, barns, and sheds. Sometimes a wall has had to be removed to extricate the homebuilt from its hatch-

This 300 HP IO-540 engine in an SX-300 has gold-plated rocker-box covers. They certainly look nice, but they don't make it go any faster.

ery. Usually a one- or two-car garage will be large enough to build all the major components, which can then be assembled outside or in a hangar at the airport. A few homebuilts have a one-piece wing that requires about a 25- or 30-foot space to assemble, so check on this sort of thing with the designer or kit manufacturer before you commit yourself to a specific design.

There are two more important things to consider before you pick out the one and only homebuilt for you: how much time do you have available and how much money do you have for this kind of project. Before you take on the commitment of completing a homebuilt, you should realize that it takes a lot of time. You need an understanding family who will give you time to work on your project or, better yet, a family who will help you on your project. Unless you have more money available than the average middle-class family, your project will put some strain on the family budget. If you build from scratch (buy a set of plans and then buy materials as you go along), the costs will be spread out over the construction period, but there will be occasional large expenses for the engine, radios, and instruments. Building from a kit will cost you more in total and in lump sums. In addition to the materials, you will pay for the trouble of gathering all the materials together for you, the prefabrication of certain parts, the shipping, and the crating. Usually, kits can be divided into separate purchases, such as a wing kit, a fuselage kit, and an empanage kit.

This lets you spread out the total outlay of funds, but

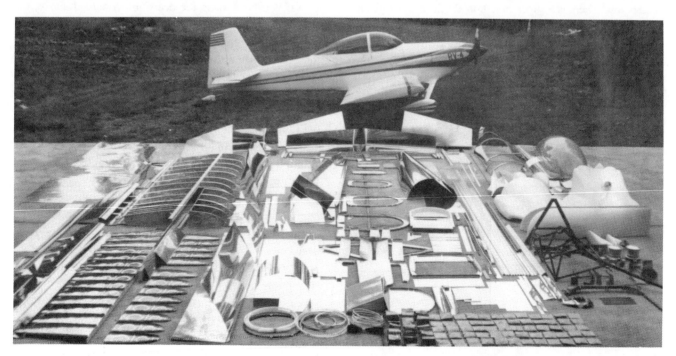

Here is what you get for $9,055 for the RV-4 kit. Add an engine, a propeller, radios, and paint, and you have an airplane (some assembly required). Photograph courtesy of Van's Aircraft, Inc.

requires a few thousand dollars at a time, as opposed to a few hundred dollars when you build from scratch. On the plus side, building from a kit greatly simplifies matters because you get all (you hope) the bits and pieces you need in a few crates. When building from scratch, you must determine from the plans or, if you are lucky, a materials list, what materials you need, find a place to buy them, and either get the materials by mail order or by numerous trips to numerous stores. If you can afford the lump-sum expenditures, a kit is the easier and quicker way to build. If not, build from scratch and you will save a considerable amount of money.

Table 2.2 lists homebuilts that can be built from plans either in kit form or from scratch. In the column headed ACRO, *mild* refers to loops and rolls and *limited* (*Lim*) means anything except snap maneuvers such as snap rolls. Costs vary greatly depending on the number of prefabricated parts used, how well you scrounge, and how elaborate your instrument panel and radio stack are. The cost estimates in the table are based on minimum equipment and a run-out engine you overhaul yourself. Where no costs figures were available, I resorted to a WAG (wild guess). If

some of the estimated "time to build" hours look too good to be true, they are probably highly optimistic figures provided by the kit manufacturer or the designer. By the time you read this, some designs may have been taken off the market, new designs may have been created, and addresses may have changed. The latest information on any design can be obtained from the designer, EAA headquarters, or possibly from your local EAA chapter.

A final word of caution: not all people who start a homebuilt project complete it. Those who do report immeasurable self-satisfaction. Consider carefully before you start. Don't take on more than you can handle moneywise, timewise, or skillwise. On the positive side, if you like making things, you will enjoy seeing a real "live" airplane taking shape under your hands. You will learn a lot as you tackle things you haven't done before, and you will meet a lot of great people in the homebuilt movement. If you don't finish, you can probably sell the project to recoup some of your expenses. If you do finish, you'll have the spectacular experience of flying an airplane you built with your own hands.

**TABLE 2.2.  Reference List of Homebuilt Aircraft**

| Name | Material | Seat | Wing/Gear | Cruise | Acro | Engine | Cost | Estimated Time (hr) | Address |
|------|----------|------|-----------|--------|------|--------|------|----------|---------|
| Ace | Wood | 1 | High/Tail | 75MPH | No | 1500cc-1800VW | $5K | 500 | Ace Aircraft 106 Authur Rd. Ashville, NC 28806 |
| Acro I | Composite | 1 | Low/Tail | 160MPH | Yes | 125-180HP | $25K | 700 | Aircraft Technologies 4265 Liburn Industrial Way Liburn, GA 30247 |
| Aerosport Quail | Metal | 1 | High/Tri | 110MPH | No | 1500cc-1800VW | $6K | 300 | Aerosport, Inc. P.O. Box 278 Holly Springs, NC 27540 |
| Aerosport Scamp | Metal | 1 | Biwing/Tri | 85MPH | Mild | 1600cc-2100VW | $6K | 500 | Same as above |
| Aerosport Woody Pusher | Wood or steel tube | 2 | Parasol/Tail | 87MPH | No | 60-85HP | $4K | 1600 | Same as above |
| Air Master | Composite | 2 | Low/Tri | 140MPH | No | 150HP | $20K | 2000 | Air Master 7822 Gulfton Houston, TX 77036 |
| Airshark I | Composite | 4 | Low/Tri amphibian | 95MPH | No | 200HP | $35K | 3000 | Composite Aircraft Design 450 Hamlin Ave. Satellite Beach, FL 32937 |
| American Eaglet | Composite kit | 1 | High/Single centerline | 52MPH | No | 12HP | $4K | 500 | AmEagle Corp. 841 Winslow Ct. Muskegon, MI 49441 |
| AMF-S14 | Wood | 2 or 3 | High/Tail | 120MPH | No | 150HP | $12K | 1500 | Falconar Aviation, Ltd. 19 Airport Rd. Edmonton, Alberta T5E 0W7 |

**TABLE 2.2.**   *(continued)*

| Name | Material | Seat | Wing/Gear | Cruise | Acro | Engine | Estimated Cost | Time (hr) | Address |
|------|----------|------|-----------|--------|------|--------|------|-----------|---------|
| ARV-1K Golden Hawk Ultralight | Composite | 2 | High/Tri & canard | 100MPH | No | 55HP | $16K | 500 | Same as above |
| Advenger Gyroplane | Aluminum tube | 2 | Rotary/Tri | 80MPH | No | 150HP | $20K | 2000 | Marchetti Engineering 38325 Primrose Spring Grove, IL 60081 |
| Avid Bandit | Metal kit | 2 | High/Tail | 75MPH | No | 65HP | $14K | 1200 | Avid Aircraft P.O. Box 728 Caldwell, ID 83606 |
| Avid Catalina | Metal kit | 3 | High/ Amphibian | 75MPH | No | 65HP | $19K | 1700 | Same as above |
| Avid Magnum | Metal kit | 2 | High/Tail | 130MPH | No | 160HP | $23K | 1500 | Same as above |
| Avid Mark IV | Metal kit | 2 | High/Tail | 110MPH | Lim | 65HP | $19K | 1500 | Same as above |
| Baby Lakes | Steel tube & wood | 1 | Biplane/ Tail | 118MPH | Yes | 60-100HP | $13K | 2000 | Barney Oldfield Aircraft Co. P.O. Box 228 Needham, MA 02192 |
| Bakeng Deuce | Steel tube & wood | 2 | Parasol/ Tail | 140MPH | No | 125-150HP | $9K | 1000 | Bakeng Aircraft 19025 92nd West Edmonds, WA 98020 |
| Barracuda | Wood | 2 | Low/Tri | 180MPH | Lim | 180-300HP | $15K | 2000 | Buethe Enterprises, Inc. P.O. Box 486 Cathedral City, CA 92234 |
| BD-4 | Metal & composite | 2 or 4 | High/Tri | 180MPH | No | 108-300HP | $10K | 900 | Bede Four Sales, Inc. P.O. Box 232 Tallmadge, OH 44278 |
| BD-5 | Metal kit | 1 | Low/Tri | 229MPH | Yes | 70HP | $15K | 800 | Same as above |
| BD-5J | Metal kit | 1 | Low/Tri | 450MPH | Yes | Jet | $45K | 1600 | Same as above |
| BD-10 | Composite kit | 2 | Mid/Tri | Mach 1.4 | Yes | CJ-610 or J-85 | $225K | 3500 | Bede Jet Corp. 18421 Edison Ave. Chesterfield, MO 63005 |
| Benson Gyrocopter | Metal | 1 | Rotary/Tri | 60MPH | No | 72-90HP | $8K | 500 | Ken Brook Mfg. 11852 Western Ave. Stanton, CA 90680 |
| Berkut | Composite | 2 | Mid/Tri & canard | 240MPH | No | 205HP | $17K | 1700 | Experimental Aviation 3025A Airport Ave. Santa Monica, CA 90405 |
| Bowers Fly Baby | Wood | 1 | Low/Tail | 115MPH | Lim | 65-85HP | $8K | 700 | Peter Bowers 10458 16th Ave. S Seattle, WA 98168 |
| Bushby Midget Mustang | Metal kit | 1 | Low/Tail | 215MPH | Lim | 125HP | $10K | 600 | Mustang Aeronautics, Inc. P.O. Box 1685 Troy, MI 48099 |
| Bushby Mustang II | Metal kit | 2 | Low/Tail | 200MPH | Lim | 135-150HP | $12K | 1800 | Same as above |
| Capella | Steel tube | 2 | High/Tail | 104MPH | No | 66HP | $14K | 350 | Flightworks 4211-C Todd Lane Austin, TX 78744 |

**TABLE 2.2.**  (*continued*)

| Name | Material | Seat | Wing/Gear | Cruise | Acro | Engine | Estimated Cost | Time (hr) | Address |
|------|----------|------|-----------|--------|------|--------|------|-----------|---------|
| Carrera 180/182 Ultralight | Aluminum tube | 1 or 2 | High/Tri or tail | 180MPH | No | 65HP | $18K | 200 | Advanced Aviation, Inc. 323 North Ivey Lane Orlando, FL 32811 |
| Cassutt Racer | Steel tube & wood | 1 | Mid/Tail | 180MPH | Yes | 85-150HP | $9K | 400 | Southern Aero. Corp. 14100 Lake Candlewood Ct. Miami Lakes, FL 33014 |
| Cavalier | Wood | 2 | Low/Tri | 155MPH | No | 85-135HP | $10K | 1800 | K & S Aircraft 4623 Fortune Rd. SE Calgary, Alberta T2A 2A7 |
| Celerity | Wood & composite | 2 | Low/Tail | 205MPH | No | 160HP | $15K | 1800 | Mirage Aircraft Corp. 3936 Austin St. Klamath Falls, OR 97603 |
| Christavia MK 1 | Steel tube | 2 | High/Tail | 105MPH | No | 65HP | $12K | 2000 | Elmwood Aviation RR 4, Elmwood Dr. Belleville, Ontario K8N 4Z4 |
| Christavia MK 4 | Steel tube | 4 | High/Tail | 120MPH | No | 150HP | $15K | 2500 | Same as above |
| Christen Eagle II | Steel tube kit | 2 | Biplane/Tail | 165MPH | Yes | 200HP | $55K | 2000 | Christen Industries, Inc. 1048 Santa Ana Valley Rd. Hollister, CA 95023 |
| Cirrus VK 30 | Composite | 5 | Low/Tri | 250MPH | No | 300HP | $60K | 4000 | Cirrus Design Corp. S3440A Highway 12 Baraboo, WI 53913 |
| Commander Gyroplane | Aluminum tube | 1 or 2 | Rotary/Tri | 45MPH | No | 40HP | $7K/ $15K | 1000 | Air Commander International 1585 Aviation Center Parkway Daytona Beach, FL 32114 |
| Corben Baby Ace | Steel tube & wood | 1 | Parasol/Tail | 85MPH | No | 65-125HP | $8K | 1000 | EAA Aviation Center P.O. Box 3086 Oshkosh, WI 54903-4800 |
| Corben Super Ace | Steel tube & wood | 1 | Parasol/Tail | 110MPH | No | 85HP | $10K | 1000 | Same as above |
| Corby Starlet | Wood | 1 | Low/Tail | 135MPH | Lim | 1500cc-1800VW | $9K | 1000 | CSN 510 NW 46th Terrace Plantation, FL 33317 |
| Cozy Mark IV | Composite | 3 | Mid/Tri | 210MPH | No | 160-200HP | $20K | 2500 | CO-Z Development 2046 N. 63rd Place Mesa, AZ 85205 |
| Cubmajor/ Majorette | Steel tube & wood | 2 | High/Tail | 110MPH | No | 100HP | $15K | 1100 | Falconar Aviation 19 Airport Rd. Edmonton, Alberta T5E 0W7 |
| Cvjetkovic CA-61 | Wood | 1 or 2 | Low/Tail | 118MPH | No | 65HP | $7K | 1000 | Anton Cvjetkovic P.O. Box 323 Newbury Park, CA 91320 |
| Cvjetkovic CA-65 | Wood | 2 | Low/Tail | 125MPH | No | 125HP | $12K | 1500 | Same as above |
| Cvjetkovic CA-65A | Metal | 2 | Low/Tail | 135MPH | Yes | 108-125HP | $15K | 2000 | Same as above |
| DA-2A | Metal | 2 | Low/Tri | 140MPH | No | 100HP | $13K | 1800 | D2, Inc. P.O. Box 524 LaPine, OR 97739 |

**TABLE 2.2.** (*continued*)

| Name | Material | Seat | Wing/Gear | Cruise | Acro | Engine | Estimated Cost | Time (hr) | Address |
|------|----------|------|-----------|--------|------|--------|------|------|---------|
| Defiant | Composite | 4 | Mid/Tri & canard | 175MPH | No | 160HP (2) | $30K | 3000 | Aircraft Spruce 201 W. Truslow Ave. Fullerton, CA 92632 |
| Dragonfly | Composite | 2 | Low/Tail & canard | 150MPH | No | 80HP | $11K | 1500 | Viking Aircraft P.O. Box 20791 Carson City, NV 89721 |
| Dyke Delta | Steel tube | 4 | Delta/Tail | 170MPH | No | 180HP | $17K | 2500 | Dyke Aircraft 2840 Old Yellow Springs Rd. Fairborn, OH 45324 |
| EAA Acro Sport | Steel tube & wood | 1 | Biplane/ Tail | 130MPH | Yes | 100HP 200HP | $12K | 2500 | EAA Aviation Center P.O. Box 3086 Oshkosh, WI 54903-4800 |
| EAA Acro Sport II | Steel tube & wood | 2 | Biplane/ Tail | 123MPH | Yes | 180HP | $14K | 2800 | Same as above |
| EAA Pober Junior Ace | Steel tube & wood | 2 | Parasol/ Tail | 85MPH | No | 85HP | $8K | 800 | Same as above |
| EAA Pober Pixie | Steel tube & wood | 1 | Parasol/ Tail | 85MPH | No | 1600cc- 2100VW | $6K | 800 | Same as above |
| Europa | Composite | 2 | Mid/Tail | 150MPH | No | 80HP | $20K | 2000 | Europa Aviation, Ltd. Kirby Mills Industrial Kirkbymoorside N. Yorkshire, UK Y06 5NR |
| Falco | Wood | 2 | Low/Tri | 200MPH | Lim | 160HP | $25K | 2500 | Sequoia Aircraft Dept. S, 2000 Tomlynn St. Richmond, VA 23230 |
| Falconar SAL Mustang | Wood, aluminum | 1 or 2 | Low/Tail | 190MPH | Lim | 175- 200HP | $25K | 2500 | Falconar Aviation, Ltd. 19 Airport Rd. Edmonton, Alberta, T5E 0W7 |
| FEW Mustang | Composite kit | 1 | Low/Tail | 200MPH | Lim | 250HP | $40K | 2500 | Fighter Escort Wings Ardmore Air Park A206 Gene Autry, OK 73436 |
| Fike Model E | Steel tube & wood | 2 | High/Tail | 80MPH | No | 60HP | $5K | 800 | W. J. Fike P.O. Box 683 Anchorage, AL 99501 |
| Fisher Aero Celebrity | Wood | 2 | Biplane/ Tail | 85MPH | No | 75HP | $15.5K | 1500 | Fisher Aero Corp. Rt 2, Box 197 Portsmouth, OH 45662 |
| Fisher Aero Horizon 2 | Wood | 2 | High/Tail | 85MPH | No | 75HP | $15.4K | 1500 | Same as above |
| Fisher Aero Horizon | Wood | 2 | High/Tail | 85MPH | No | 75HP | $13.7K | 1500 | Same as above |
| Fisher Dakota Hawk | Wood | 2 | High/Tail | 85MPH | No | 85-95HP | $14K | 500 | Same as above |
| Firestar Ultralight | Metal | 1 | High/Tail | 70MPH | No | Rotax 277 | $7K | 800 | KOLB Co. R.D. 3, Box 38 Phoenixville, PA 19460 |

**TABLE 2.2.** (*continued*)

| Name | Material | Seat | Wing/Gear | Cruise | Acro | Engine | Estimated Cost | Time (hr) | Address |
|------|----------|------|-----------|--------|------|--------|------|-----------|---------|
| Firestar II Ultralight | Metal | 2 | High/Tail | 85MPH | No | Rotax 447 | $8K | 800 | Same as above |
| Flying Flea | Wood | 2 | Parasol/ Tail | 95MPH | No | 90HP | $8K | 1000 | Falconar Aviation, Ltd. 19 Airport Rd. Edmonton, Alberta T5E 0W7 |
| Genesis I Sailplane | Composite | 1 | Mid/Center | 130MPH | No | N/A | $10K | 300 | Group Genesis, Inc. 1530 Pole Lane Rd. Marion, OH 43302 |
| Glasair II-S | Composite kit | 2 | Low/Tri | 235MPH | Lim | 180HP | $35K | 3000 | Stoddard-Hamilton Aircraft 18701 58th Ave, NE Arlington, WA 98223 |
| Glasair III | Composite | 2 | Low/Tri | 300MPH | Lim | 300HP | $45K | 3200 | Same as above |
| Gla Star | Composite | 2 | High/Tri | 140MPH | No | 125HP | $20K | 2000 | Same as above |
| GT-500 Quicksilver | Aluminum | 2 | High/Tri | 45MPH | No | Rotax 582 | $18K | 150 | Quicksilver Enterprises 27495 Diaz Rd. Temecula, CA 92590 |
| Hartz CB-1 | Steel tube & wood | 2 | Biplane/ Tail | 100MPH | No | 90-150HP | $10K | 1800 | Dudley Kelly Rt 4 Versailles, KY 40383 |
| Hawk I Gyroplane | Metal kit | 2 | Rotary/Tri | 65MPH | No | 180HP | $12K | 900 | Groen Brothers Aviation 1784 W. 700 South Salt Lake City, UT 84104 |
| Hawk II Ultralight | Aluminum tube | 1 or 2 | High/Tail | 75MPH | No | 52HP | $12K | 200 | CGS Aviation P.O. Box 41007 Brecksville, OH 44141 |
| Hawker Hurricane | Wood | 1 | Low/Tail | 185MPH | Yes | 150HP | $14K | 2000 | Sindlinger Aircraft 5923 9th St. NW Puyallup, WA 98371 |
| Hummel Bird | Metal | 1 | Low/Tail | 105MPH | No | 30HP | $7K | 900 | Morry Hummel 509 East Butler Bryan, OH 43506 |
| Humming-bird | Metal & composite | 4 | Rotary/ Tri | 90MPH | No | 245HP | $70K | 3500 | Vertical Aviation Tech. P.O. Box 2527 Sanford, FL 32772-2527 |
| Hurricane Gyroplane | Aluminum & composite | 1 | Rotary/ Tri | 85MPH | No | 65HP | $20K | 3000 | Wind Ryder Engineering 555 Alter #15 Bloomfield, CO 80020 |
| J-1B Don Quixote | Wood | 1 | High/Tail | 75MPH | No | 50HP | $5K | 1500 | Alpha Aviation Supply Co. P.O. Box 8641 Greenville, TX 75401 |
| Javelin Sport Racer | Steel tube & wood | 2 | Mid/Tail | 175MPH | Lim | 200HP | $12K | 1800 | Sport Racer, Inc. 14 Hawthorne Rd. Valley Center, KS 67147 |
| Javelin Wichawk | Steel tube & wood | 2 | Biplane/ Tail | 135MPH | Mild | 125-300HP | $10K | 2000 | Javelin Aircraft Municipal Airport Augusta, KS 67010 |
| Javelin V6 STOL | Steel tube | 4 | High/Tail | 120MPH | No | 230HP | $15K | 1800 | Same as above |

**TABLE 2.2.** *(continued)*

| Name | Material | Seat | Wing/Gear | Cruise | Acro | Engine | Estimated Cost | Time (hr) | Address |
|------|----------|------|-----------|--------|------|--------|------|------|---------|
| Jodel F-9 & F-10 | Wood | 1 | Low/Tail | 100MPH | No | 60HP | $6K | 700 | Falconar Aviation, Ltd. 19 Airport Rd. Edmonton, Alberta T5E 0W7 |
| Jodel F-11 | Wood | 2 | Low/Tail | 125 MPH | No | 100HP | $10K | 1200 | Same as above |
| Jodel F-12 | Wood | 3 | Low/Tail | 140MPH | No | 140HP | $13K | 1500 | Same as above |
| Jenny | Steel tube & aluminum | 2 | Biplane/ Tail | 60MPH | No | 46HP | $12K | 1500 | Leading Edge Airfoils 331 South 14th St. Colorado Springs, CO 80904 |
| Jungster I | Wood | 1 | Biplane/ Tail | 119MPH | Yes | 85-150HP | $11K | 650 | K & S Aircraft 4623 Fortune Rd. SE Calgary, Alberta T2A 2A7 |
| Jungster II | Wood | 1 | Parasol/ Tail | 148MPH | Yes | 85-180HP | $11K | 600 | Same as above |
| KB-3 Gyrocopter | Metal | 1 | Rotary/ Skid | 60MPH | No | Rotax 532 | $10K | 1000 | Ken Brook Manufacturing 11852 Western Ave. Stanton, CA 90680 |
| KIS | Composite kit | 2 | Low/Tail or Tri | 165MPH 135MPH | No | 180HP 80HP | $11K | 1500 | Tri-R Technologies, Inc. 1114 E. 5th St. Oxnard, CA 93030 |
| Kitfox Speedster | Metal kit | 2 | High/Tail | 125MPH | Mild | Rotax 912 | $12K | 1200 | Skystar Aircraft Corp. 100-AS N. Kings Rd. Nampa, ID 83687 |
| Kitfox IV | Metal kit | 2 | High/Tail | 75MPH | No | 52-80HP | $10K | 1000 | Same as above |
| Kitfox XL | Metal kit | 2 | High/Tail | 75MPH | No | 52-80HP | $12K | 1000 | Same as above |
| Kitfox Vixen | Metal kit | 2 | High/Tail | 125MPH | Mild | Rotax 912 | $14K | 1200 | Same as above |
| Kit Hawk | Metal | 2 | Biplane/ Center | 100MPH | No | 65HP | $20K | 2000 | Kestrel Sport Aviation P.O. Box 1808 Brockville, Ontario K6V 6K6 |
| KR-1 & KR-2 | Wood & composite | 1 2 | Low/Tail | 130MPH | No | 1700cc VW | $8K | 750 | Rand Robinson Engr, Inc. 15641 Product Lane Suite A-5 Huntington Beach, CA 92649 |
| Lancair IV & ES | Composite kit | 4 | Low/Tri | 190MPH/ 330MPH | Lim | 200-250HP | $45K | 2500 | Neico Aviation 2244 Airport Way Redmond, OR 97756 |
| Lancair 320/360 | Composite kit | 2 | Low/Tri | 240MPH | Lim | 160-180HP | $40K | 2500 | Same as above |
| LM-1 & LM-2 | Aluminum tube | 1 | High/Tail | 65MPH | No | 35-40HP | $7K | 900 | Light Minature Aircraft Building 411 Opa-Locka A.P. Opa-Locka, FL 33054 |
| Loehle 51-51 | Wood kit | 1 | Low/Tail | 80MPH | No | 50HP | $8K | 1000 | Loehle Aviation, Inc. Shipmans Creek Rd. Wartrace, TN 37183 |
| Lonestar | Steel tube | 1 | Rotary/ | 65MPH | No | 64HP | $19K | 1000 | Star Aviation, Inc. |

**TABLE 2.2.** *(continued)*

| Name | Material | Seat | Wing/Gear | Cruise | Acro | Engine | Cost | Estimated Time (hr) | Address |
|------|----------|------|-----------|--------|------|--------|------|---------------------|---------|
| Sport Helicopter | | | Skid | | | | | | 20180 FM 2252 San Antonio, TX 78226-2610 |
| Long-EZ | Composite | 2 | Mid/Tri & canard | 175MPH | No | 108HP | $15K | 1500 | Aircraft Spruce 201 W. Truslow Fullerton, CA 92632 |
| Mariner | Metal | 1 or 2 | Biplane/ Tail/ Amphibian | 55MPH | No | 40HP | $11K | 1600 | Two Wings Aviation 6821 167th Ave. Forest Lake, MI 55025 |
| Merlin | Metal | 2 | High/Tail | 140MPH | No | 150HP | $20K | 2000 | Merlin Aircraft 509 Airport Rd. Hangar 1C Muskegon, MI 49441 |
| Meyer's Little Toot | Steel tube & wood | 1 | Biplane/ Tail | 110MPH | Yes | 90-200HP | $12K | 2500 | Meyer Aircraft 5706 Abby Dr. Corpus Christi, TX 78413 |
| Micro Mong | Steel tube & aluminum | 1 | Biwing/ Tail | 95MPH | Mild | Rotax 503 | $8K | 1200 | Ed Fisher 2331 Dodgeville Rd. Rome, OH 44085 |
| MINI-MAX | Wood | 1 | Mid or High/Tail | 85MPH | No | 50-65HP | $8K | 900 | Team, Inc. Rt 1, Box 338 Bradyville, TN 37026 |
| Mini-500 | Aluminum & steel tube | 1 | Rotary/ Skid | 75MPH | No | 67HP | $22K | 2000 | Revolution Helicopter Corp. 1905 W. Jesse James Rd. Excelsior Springs, MO 64024 |
| Mitchell Wing Ultralight | Wood & composite | 1 | High/Tail | 34MPH | Mild | 12HP | $4.5K | 200 | M Company 1900 S. Newcomb Porterville, CA 93257 |
| Monnett Sonerai | Metal | 1 | Mid/Tail | 150MPH | Yes | 1600cc VW | $9K | 800 | Great Plains Aircraft Supply P.O. Box 304 St. Charles, IL 60174 |
| Mooney Mite | Wood kit | 1 | Low/Tri | 130MPH | No | 65HP | $8K | 500 | Mooney Aircraft Co. P.O. Box 3999 Charlottesville, VA 22903 |
| Morrisey Bravo | Metal kit | 2 | Low/Tri | 145MPH | Lim | 150HP | $20K | 2000 | Morrisey Aircraft P.O. Box 4129 Oceanside, CA 92054 |
| Murphy Rebel | Metal kit | 2 | High/Tail | 110MPH | No | 115HP | $16K | 1500 | Murphy Aircraft Dept C, 8880C Young Rd. S. Chilliwack, B.C. V2P 4P5 |
| Murphy Renegade | Metal kit | 2 | Biplane/ Tail | 80MPH | No | 65HP | $12K | 1500 | Same as above |
| Nieuport 11 or 12 | Metal | 2 | Biplane/ Tail | 65MPH | No | 46HP | $9K | 900 | Leading Edge Airfoils 331 South 14th St. Colorado Springs, CO 80904 |
| Nuwaco T-10 | Steel tube | 3 | Biplane/ Tail | 130MPH | No | 275HP | $60K | 2000 | Aircraft Dynamics, Inc. 2978 E. Euclid Place Littleton, CO 80121 |
| OMEGA II | Metal kit | 2 | Low/Tri | 200MPH | No | 200HP | $40K | 2200 | Systems Aero Engineering 1850 N. 600 West Logan, UT 84321 |

**TABLE 2.2.** (*continued*)

| Name | Material | Seat | Wing/Gear | Cruise | Acro | Engine | Estimated Cost | Time (hr) | Address |
|------|----------|------|-----------|--------|------|--------|------|-----------|---------|
| Osprey II | Wood | 2 | Mid/ Amphibian | 130MPH | No | 150HP | $20K | 2000 | Osprey Aircraft 3741 El Ricon Way Sacramento, CA 95825 |
| Pazmany PL-2 | Metal | 2 | Low/Tri | 136MPH | Lim | 109-150HP | $20K | 3500 | Pazmany Aircraft Corp. P.O. Box 80051 San Diego, CA 92138 |
| Pazmany PL-4A | Metal | 1 | Low/Tail | 100MPH | No | 1600cc VW | $9K | 1500 | Pazmany Aircraft Corp. P.O. Box 80051 San Diego, CA 92138 |
| PDQ-2C Ultralight | Aluminum tube | 1 | Mid/Tri | 70MPH | No | 50HP | $5K | 500 | PDQ Aircraft Products 28975 Alpine Lane Elkhart, IN 46514 |
| Pegazair | Metal | 2 | High/Tail | 115MPH | No | 65-100HP | $10K | 1000 | Pegase Aero 437 Route 309 Nord Mont-St-Michel, Quebec J0W 1P0 |
| Pelican Club | Metal & composite | 2 | High/ Floats | 105MPH | No | 80HP | $16K | 1500 | Ultravia Aero, Inc. 300 D Airport Rd. Mascoche, Quebec J7K 3C1 |
| Phoenix L-II | Metal | 2 | High/Tri | 130MPH | No | 100-180HP | $17K | 1800 | Leeward Aeronautical 890 NE 149th St. Miami, FL 33161 |
| Pietenpol Air Camper | Wood | 2 | Parasol/ Tail | 90MPH | No | 65HP | $7K | 1200 | John W. Grega 355 Grand Blvd. Bedford, OH 44146 |
| Pitts S-1 | Steel tube & wood | 1 | Biplane/ Tail | 140MPH | Yes | 125-180HP | $20K | 2000 | Pitts Aeronautics P.O. Box 547 Afton, WY 83110 |
| Pitts S-2A | Steel tube & wood | 2 | Biplane/ Tail | 152MPH | Yes | 200HP | $25K | 2200 | Same as above |
| Pulsar | Composite kit | 2 | Low/Tri | 140MPH | No | 66-80HP | $26K | 2500 | Aero Designs, Inc. 11910 Radium St. San Antonio, TX 78216 |
| Q-2 | Composite | 2 | Mid/Tri | 170MPH | No | 100HP | $10K | 1000 | See Trade-A-Plane ads |
| Q-200 | Composite kit | 2 | Shoulder | 180MPH | No | 100HP | $9K | 900 | See Trade-A-Plane ads |
| Questair Spirit | Metal | 2 | Low/Tri | 190MPH | Lim | 200HP | $17K | 1200 | Questair, Inc. 3930 Ventura Dr. Suite 450 Arlington Heights, IL 60004 |
| Questair Venture | Metal | 2 | Low/Tri | 240MPH | Lim | 260HP | $20K | 1500 | Same as above |
| Quickie | Wood & composite | 1 | Mid/Tail & canard | 120MPH | No | 16-18HP | $6K | 400 | See Trade-A-Plane ads |
| Quick-silver | Metal kit | 2 | High/Tri | 70MPH | No | 50-60HP | $18K | 200 | Quicksilver Enterprises P.O. Box 1572 Temecula, CA 92390 |
| Rans S-4 Coyote | Aluminum tube | 1 | High/Tri | 60MPH | No | 42HP | $9K | 900 | Rans Company 4600 Highway 183 Alt. Hays, KS 67601 |

**TABLE 2.2.** (*continued*)

| Name | Material | Seat | Wing/Gear | Cruise | Acro | Engine | Estimated Cost | Time (hr) | Address |
|------|----------|------|-----------|--------|------|--------|----------------|-----------|---------|
| Rans S-7 Courier | Metal kit | 2 | High/Tail | 70MPH | No | 65HP | $8K | 900 | Same as above |
| Rans S-9 Chaos | Steel tube | 1 | Mid/Tail | 100MPH | No | 48HP | $12K | 1500 | Same as above |
| Rans S-10 Sakota | Steel tube | 2 | Mid/Tail | 95MPH | No | 65HP | $13K | 1800 | Same as above |
| Rans S-6 Coyote II | Aluminum tube | 2 | High/Tri | 80MPH | No | 65HP | $16K | 1200 | Same as above |
| Redfern Fokker Triplane | Steel tube & wood | 1 | Triplane/ Tail | 100MPH | No | 145HP | $15K | 2500 | Walt Redfern S-211 Spencer Post Falls, ID 83854 |
| Redfern Nieuport 17 | Steel tube & wood | 1 | Biplane/ Tail | 110MPH | No | 145HP | $14K | 2500 | Same as above |
| RLU Breezy | Steel tube | 2 | Parasol/ Tri | 75MPH | No | 90HP | $6K | 800 | Charles B. Roloff P.O. Box 358 Palos Park, IL 60464 |
| Rogers Sportaire | Steel tube & wood | 2 | Low/Tri | 132MPH | No | 125HP | $12K | 2000 | Rogers Aircraft Co. 758 Libby Dr. Riverside, CA 92507 |
| Rotor Lightning Sport Copter | Steel tube | 1 | Rotary/ Tri | 50MPH | No | 46HP | $12K | 1800 | Vaneraft Copters 7246 N. Mohawk Portland, OR 97203 |
| Rotorway Scorpion Too | Steel tube & composite kit | 2 | Rotary/ Skid | 80MPH | No | 133HP | $25K | 2500 | Rotorway International 4141 W. Chandler Blvd. Chandler, Az 85226 |
| Rotorway Executive | Steel tube & composite kit | 2 | Rotary/ Skid | 80MPH | No | 133HP | $40K | 3000 | Same as above |
| S-51D | Metal | 1 or 2 | Low/Tail | 235MPH | Yes | 350HP | $45K | 3000 | Stewart 51, Inc. Vero Beach Airport P.O. Box 6070 Vero Beach, FL 32961 |
| Sea Hawk/ Glass Goose | Composite kit | 2 | Biplane/ Amphibian | 140MPH | No | 160HP | $20K | 2000 | Quickit, Inc. 9002 Summer Glen Dallas, TX 75243 |
| Seawind | Composite | 4 | Mid/Tri amphibian | 190MPH | No | 300HP | $55K | 300 | S.N.A., Inc. Box 607 Kimberton, PA 19442-0607 |
| SE-5A | Wood | 1 | Biplane/ Tail | 90MPH | No | 65-100HP | $8K | 1200 | Replica Plans 9531 Kirkmond Rd. Richmond, B.C. V7E 1M7 |
| Sequoia | Steel tube & composite | 2 | Low/Tri | 217MPH | Yes | 320HP | $17K | 2500 | Sequoia Aircraft Corp. 900 West Franklin St. Richmond, VA 23220 |
| Smith Miniplane | Steel tube & wood | 1 | Biplane/ Tail | 122MPH | Lim | 65-125HP | $8K | 2000 | Dorthy Smith & Son 3502 Sunny Hills Dr. Norco, CA 91760 |
| Smyth Sidewinder | Metal | 2 | Low/Tri | 170MPH | Lim | 90-180HP | $10K | 2000 | Smyth Aerodynamics P.O. Box 308 Huntington, IN 46750 |

**TABLE 2.2.**   (*continued*)

| Name | Material | Seat | Wing/Gear | Cruise | Acro | Engine | Estimated Cost | Time (hr) | Address |
|------|----------|------|-----------|--------|------|--------|------|-----------|---------|
| Sorrell Hiperbipe | Steel tube & wood | 2 | Biplane/Tail | 160MPH | Yes | 180HP | $10K | 2000 | Sorrell Aviation Box 660, Rt 1 Tenino, WA 98589 |
| Space-walker | Steel tube & wood | 1 | Low/Tail | 150MPH | Yes | 65-115HP | $8K | 1700 | Hirt Aircraft Corp. P.O. Box 2134 Hemet, CA 92546 |
| Speedtwin | Metal | 2 | Low/Tail | 173MPH | Lim | 100HP (2) | $18K | 2000 | Speedtwin Developments Upper Cae Garw Farm Trellech, Monmouth, Gwent NP54PJ, U.K. (0600)860165 |
| Spencer Air Car | Wood & composite | 4 | High/Tri amphibian | 135MPH | No | 260HP | $30K | 3500 | Spencer Air Car 8725 Oland Ave. Sun Valley, CA 91352 |
| Steen Skybolt | Steel tube & wood | 2 | Biplane/Tail | 130MPH | Yes | 180-250HP | $13K | 2000 | Steen Aero Lab, Inc. 1210 Airport Rd. Marion, NC 28752 |
| Stewart Headwind | Steel tube & wood | 1 | High/Tail | 75MPH | No | 36-85HP | $6K | 1000 | Stewart Aircraft Corp. 11420 State Route 165 Salem, OH 44460 |
| Stewart Foo Fighter | Steel tube & wood | 1 | Biplane/Tail | 115MPH | No | 130HP | $8K | 1500 | Same as above |
| Stits Playboy | Wood | 1 | Low/Tail | 135 MPH | No | 65-125HP | $8K | 1000 | See Trade-A-Plane ads |
| Stolp Starduster | Steel tube & wood | 1 | Biplane/Tail | 160MPH | No | 125HP | $13K | 1500 | Stolp Starduster Corp. 4301 Twining Riverside, CA 92509 |
| Stolp Starduster Too | Steel tube & wood | 2 | Biplane/Tail | 150MPH | Lim | 180-250HP | $15K | 2000 | Same as above |
| Stolp Acroduster | Steel tube & wood | 2 | Biplane/Tail | 150MPH | Lim | 250HP | $18K | 2200 | Same as above |
| Streak Shadow | Wood & composite | 2 | High/Tri | 100MPH | No | 65HP | $16K | 1000 | Laron Aviation Tech. R.R. 1, Box 69B Old Municipal Airport Portales, NM 88130 |
| Supercat | Wood | 1 | Low/Tail | 70MPH | No | 40HP | $5K | 800 | Wicks Aircraft Supply 410 Pine St. Highland, IL 62249 |
| Super Emeraude | Metal | 2 | Low/Tail | 133MPH | Yes | 100-150HP | $20K | 2000 | Claude Piel c/o E. Littner C.P. 272 Saint-Laurent, Quebec H4L 4V6 |
| Taylor Titch | Wood | 1 | Low/Tail | 155MPH | Yes | 40-90HP | $9K | 1200 | Mrs. John F. Taylor 25 Chesterfield Crescent Leigh-On-The-Sea Essex, UK |
| Taylor Monoplane | Wood | 1 | Low/Tail | 90MPH | No | 30-60HP | $7K | 1000 | Same as above |
| Taylor Coot | Wood & composite | 2 | Mid/Tri/Amphibian | 125MPH | No | 120-180HP | $20K | 2000 | Same as above |

**TABLE 2.2.** (*continued*)

| Name | Material | Seat | Wing/Gear | Cruise | Acro | Engine | Cost | Estimated Time (hr) | Address |
|------|----------|------|-----------|--------|------|--------|------|---------------------|---------|
| Taylor Mini-Imp | Metal | 1 | Shoulder/ Tri | 150MPH | Lim | 60-115HP | $8K | 700 | Same as above |
| Teenie Two | Metal | 1 | Low/Tri | 110MPH | No | 53-65HP | $5K | 500 | C. Y. Parker P.O. Box 181 Dradoon, AZ 85609 |
| Thorp T-18 | Metal | 2 | Low/Tail | 175MPH | Lim | 135-200HP | $18K | 2200 | Sport Aircraft 44211 Yucca Unit A Lancaster, CA 93534 |
| Thunder Mustang | Composite | 1 | Low/Tail | 300MPH | Yes | 650HP | $50K | 3500 | Papa Corporation 102 North Kings Rd. Nampa, ID 83687 |
| Thurston Trojan | Metal | 4 | Shoulder/ Tri/ Amphibian | 130MPH | No | 160-250HP | $20K | 2000 | David B. Thurston 169 Coleman Ave. Elmira, NY 14905 |
| Trio | Composite kit | 1 | Low/Tri | 150MPH | No | 115HP | $8K | 1700 | Hirt Aircraft Corp. P.O. Box 2134 Hemit, CA 92546 |
| Turner T-40A | Wood | 2 | Low/Tri | 150MPH | No | 60-100HP | $12K | 2700 | Turner Development 3717 Ruth Rd. Ft. Worth, TX 76118 |
| Twinstar Mark III | Metal | 2 | High/Tail | 95MPH | No | Rotax 582 | $9K | 800 | Kolb Company Rt. 3, Box 38 Phoenixville, PA 19460 |
| Twinstar Mark III Lite | Metal | 2 | High/Tail | 90MPH | No | Rotax 503 | $9K | 800 | Same as above |
| Ultra Pup | Wood & composite | 1 | High/Tail | 75MPH | No | 50HP | $5K | 500 | Preceptor Aircraft Corp. 1230 Shepard St. Hendersonville, NC 28792 |
| Van's RV-3 | Metal kit | 1 | Low/Tail | 200MPH | Lim | 125-160HP | $13K | 1200 | Van's Aircraft, Inc P.O. Box 160 North Plains, OR 97133 |
| Van's RV-4 | Metal kit | 2 | Low/Tail | 190MPH | Lim | 150-160HP | $16K | 2000 | Same as above |
| Van's RV-6 | Metal kit | 2 | Low/Tail | 190MPH | Lim | 150-160HP | $17K | 2000 | Same as above |
| VariEZE | Composite | 2 | Mid/Tri & Canard | 165MPH | No | 65-100HP | $10K | 500 | See Trade-A-Plane ads |
| Vari-Viggen | Metal & wood | 2 | Low/Tri & Canard | 150MPH | No | 150HP | $12K | 1700 | See Trade-A-Plane ads |
| Velocity | Composite kit | 4 | Mid/Tri & Canard | 200MPH | No | 200HP | $30K | 2500 | Velocity Aircraft 200 West Airport Rd. Sebastian, FL 32958 |
| VP-1 & VP-2 Volksplane | Wood | 1 or 2 | Low/Tail | 75MPH | No | 40-60HP | $5K | 700 | Evans Aircraft P.O. Box 744 La Jolla, CA 92037 |
| Volmer Sportsman | Wood | 2 | High/Tail Amphibian | 85MPH | No | 85-100HP | $18K | 2000 | Volmer Aircraft P.O. Box 5222 Glendale, CA 91201 |

**TABLE 2.2.** *(continued)*

| Name | Material | Seat | Wing/Gear | Cruise | Acro | Engine | Estimated Cost | Time (hr) | Address |
|------|----------|------|-----------|--------|------|--------|------|-----------|---------|
| Volmer Sun Fun | Aluminum tube | 1 | High/Tail | 28MPH | No | 15HP | $4K | 300 | Same as above |
| Wag-Aero CUBy | Steel tube & wood | 2 | High/Tail | 94MPH | No | 65-150HP | $15K | 1100 | Wag-Aero, Inc. P.O. Box 181 Lyons, WI 53148 |
| Wag-Aero Sportsman 2+2 | Steel tube & wood | 4 | High/Tail | 124MPH | No | 125-200HP | $20K | 2000 | Same as above |
| War Aircraft Replicas | Wood | 1 | Low/Tail | 145MPH | Lim | 65-125HP | $19K | 1500 | War Aircraft Replicas, Inc. 348 S. Eighth St. Santa Paula, CA 93060 |
| Wheeler Express | Composite kit | 4 | Low/Tri | 220MPH | No | 200-300HP | $40K | 3000 | EDI P.O. Box 609 Redmond, OR 97756 |
| White Lightning | Composite kit | 4 | Mid/Tri | 265MPH | No | 210HP | $45K | 3500 | White Lightning Aircraft P.O. Box 497 Walterboro, SC 29488 |
| Whittman Tailwind | Steel tube & wood | 2 | High/Tail | 150MPH | Lim | 85-140HP | $18K | 1800 | S. J. Wittman Red Oak Ct, Box 2672-1265 Oshkosh, WI 54903-1265 |
| Zenair CH-701 | Metal kit | 2 | High/Tri | 80MPH | No | Rotax 532 | $8K | 1200 | Zenith Aircraft Co. Mexico Memorial Airport Mexico, MO 65265-0650 |
| Zenair CH-100 | Metal kit | 1 | Low/Tri | 135MPH | Lim | 100HP | $20K | 1800 | Same as above |
| Zenair CH-200 | Metal kit | 2 | Low/Tri | 145MPH | Lim | 90-160HP | $22K | 2000 | Same as above |
| Zenair CH-300 | Metal kit | 3 | Low/Tri | 147MPH | Lim | 160HP | $25K | 2200 | Same as above |
| Zenair Zodiac CH-601 | Metal kit | 2 | Low/Tail or Tri | 140MPH | Lim | 90-160HP | $25K | 2000 | Same as above |
| Zenair STOL CH-701 | Metal kit | 2 | High/Tri | 85MPH | No | 80HP | $15K | 1800 | Same as above |
| Zephyr | Wood | 1 | Low/Tail | 150MPH | Lim | 100HP | $10K | 1800 | E. Littner 140 Philippe Goulet Repentigny, Quebec J5Y 3M1 |

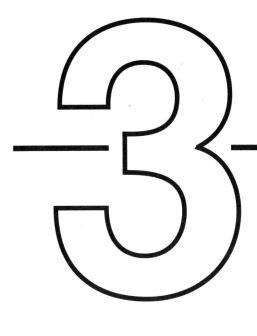

# 3 Getting Started

**LL RIGHT, you've decided on the type of plane you want to build. You've narrowed the list in Table 2.2 down to three or four possibilities. You've sent off for the information packages and talked to people who are knowledgeable about these particular homebuilts. If you're lucky, you've seen the finished products and maybe had a ride in some at a nearby EAA fly-in or open house. After looking at all the pros and cons, checking your priority list again and again, making a few compromises, and flipping a coin, you have bitten the bullet and made your final decision.**

Now comes the fun part—really getting started. Let's assume that you have picked a plane to be built from scratch rather than a kit plane. First, you write your check and send for the plans and builder's manual. Then you wait, and wait, and wait some more. You are chomping at the bit, and it seems as if it takes forever for your precious plans to arrive; but it only seems that way. Don't waste that time while you are waiting. Get your shop set up. Get organized. Get ready to start.

Table 2.1 gives you an idea of the tools you will need to add to your inventory for your particular type of construction. Buy the basic tools you need and read those operating instructions, especially the safety instructions. You don't want to lose an eye or a finger because you didn't operate the tool properly and safely. Be sure to use safety goggles or a face shield when chips or filings are flying around your workshop. Unplug power tools when not in use if chil-

dren are around. Practice with new tools on scrap material until you feel comfortable with their operation. If you are going to be working with a type of construction that is new to you, now is the time to get some training from your local community college, trade school, or a friend experienced in this skill. Some of the larger EAA open house events have forums that allow you to try your hand at riveting, welding, rib building, and fiberglass layups. Remember, welding, sheet metal working, fiberglass layups, and woodworking are like anything else; the more you practice, the better you are. So practice, practice, practice!

If you need a worktable or a workbench that you don't already have, now is the time to build it. Make sure it is level and sturdy. You may be able to just put ¾-inch plywood on your shop floor and shim it level with cardboard or thin pieces of wood and construct your large assemblies right on the plywood. This is a little harder on the knees and back,

How is this for a lean, mean, flying machine? A highly modified Cassutt Racer.

but it eliminates the construction of a large worktable. Smaller parts, such as wing ribs and tail surfaces, can usually be constructed on any handy table or a couple of strong sawhorses with a ¾-inch sheet of plywood over them. If you are going to be welding, get some firebrick on which to weld your small assemblies. Don't use regular bricks or concrete blocks. They may come apart with explosive force when heated.

Here is a homemade welding table with firebrick on top.

When at last your plans arrive, look them over carefully. Make sure all the pages are there. The plans may look formidable and complicated at first, but as you study the construction of each part in detail, they will begin to make sense. Look over your builder's guide, if one is provided, so that you are familiar with its contents. Again, check for missing or misprinted pages. If there is a suggested sequence of construction, it is advisable, but not mandatory, that you follow it. If there is no suggested sequence, it is a good idea to start out on small assemblies, such as wing ribs, the tail section, or control surfaces, and then graduate to the larger assemblies.

If there is a materials list included, you are truly a lucky duck. You can readily see what is needed for the component on which you intend to start, and you can go shopping for these goodies. If your luck is not so good and no materials list is available, you will have to study the plans and make a list of your own. Whether the materials are wood, steel, or aluminum; their dimensions should be specified on the plans. Examples include ¹⁄₁₆-inch 2024 T-3 aluminum, ⅛-inch birch ply, or ¾ × ¾-inch spruce. Add up the lengths and estimate the flat-plate areas you will need, and then add a little fudge factor. It is better to have a little material left over, which you may be able to use later, then to

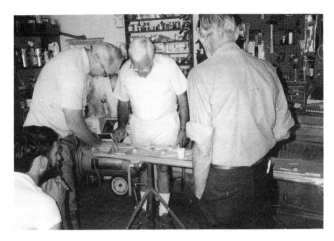

**A rib-building demonstration at an EAA meeting.**

come up a little short and be forced to make another trip to the supplier. The steel tubing and flat-plate steel most often used is 4130 chromoly for its strength and weldability. The usual aluminum material is 2024 T-3, but other aluminum alloys may be used in certain applications. Wood construction usually consists of sitka spruce and birch or mahogany plywood. Fiberglass cloth varies in weight and is normally specified for each application, as is the number of plys (layers) of cloth required.

By now, you probably know where the local sources of material are located. If not, check with your local EAA chapter or with other builders. You must be sure that the material you purchase is aircraft-quality material. After all, your life will depend on the quality of this material when you take your plane up into the wild blue.

Not all wood is created equal. What you use for wall studs in a house will not be good enough for your plane. How do you know that what the sales clerk is handing you is satisfactory? If you buy from one of the well-established supply houses, you can be 99 percent sure the material you get is airworthy. Ultimately, though, it is your responsibility to ensure that everything that goes into your bird is of good quality. Appendix A at the back of this book provides a partial list of companies that provide quality parts and materials. If you can't get everything you need locally, you will have to order by mail. Your local EAA members will be able to show you catalogs from various mail order supply houses, and you can order a catalog for yourself with your first order. As the builder of this magnificent machine, you can use whichever parts and materials you wish to use, but you should know what you are doing if you deviate from the designer's specifications. You may find parts in your local auto parts store, materials in your local lumberyard, or

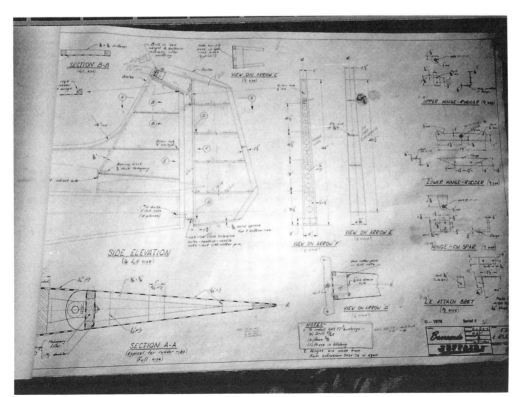

Plans come in all shapes and sizes. These are for the Barracuda.

bits and pieces in an aircraft salvage yard, but always use caution and good judgment when making changes or substitutions. When in doubt, check with the designer, a knowledgeable builder, your local EAA chapter, or EAA headquarters.

If you are ordering a kit, you won't have to search high and low for the materials you need. There may be certain finishing items, such as paint, that will not be included, but the construction materials, hardware, and supplies should all be provided in the kit. Even the crating material should be saved to use as jigs, frames, and supports. It will be your job to shape these raw materials into the vital components of your dream plane. High-cost items such as engines, radios, and instruments will probably not be included with the kit, so you will get your chance to shop around when you are ready to purchase these parts.

Now let's take a moment to look back and see where we have been and where we are going. You have analyzed your personal requirements and have studied the available homebuilts on the market. You have selected the design that best fits your desires, your time constraints, and your pocketbook. You have ordered the plans or kit and have learned (or started learning) any special skills required for the type of construction on which you are embarking. You have prepared your workshop and secured the tools you need to start fabricating parts. It looks as if the time has come to start making the sawdust fly, the aluminum bend, the welding torch burn, or the fiberglass resin flow. It's time to start building!

This stack of catalogs gives you an idea of how many things you can order by mail.

This could be your home-built aircraft if you jump in with both feet and just start building.

# 4 Time Flies When You're Having Fun

**T**IME—THE CLOCK KEEPS TICKING. As the song says, "Another day older and deeper in debt." You started off with a flurry of activity, cutting and assembling pieces, putting small assemblies together into larger assemblies, and forming major components. The fuselage has taken shape. The tail assembly and wings have come into being. A glance at the calendar shows that a great number of months have slipped by since you ordered your plans or your kit. A glance at the tasks left to complete shows a vast number of systems to tackle.

All these involve disciplines with which you may not be intimately familiar: an electrical system, a flight control system, a fuel system, possibly a hydraulic system, engine installation, an engine control system, instruments, a pitot static system, radios, antennas, brakes—the list seems to be endless. But never fear, there is an end to the list, and you can get through each and every system and each and every task if you take them one step at a time and one task at a time.

Now is a good time to regroup. If your plans or your builder's manual has a construction sequence, then by all means follow it. If not, you need to develop a sequence for yourself. During the construction phase of the primary structure, you may have "put the cart in front of the horse" on occasion and had to tear something out so that you could get something else in. Obviously, you must build and hang the engine mount before you hang the engine, and you must hang the engine before you connect the engine controls. But some horse/cart sequences may not be so obvious and will require some forethought and planning.

Moreover, you may be able to combine work on more than one system to save time and effort. For instance, when finishing the tail section, you may have a position light, a trim tab motor, and a radio antenna built into the tail as-

The builder of this MINI-MAX takes time out for a photo-op.

sembly. It would be advantageous to run the electrical wires and antenna cable to the cockpit area at the same time. The sequence of tasks may look something like this:

- Install taillight fixture
- Measure wires for taillight and connect to fixture
- Install radio antenna
- Measure coaxial cable and connect to antenna
- Install elevator trim motor

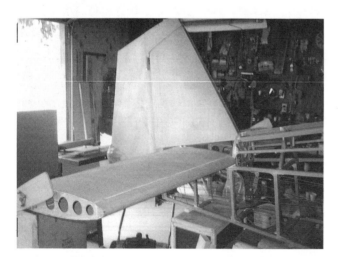

Barracuda tail feathers with positionlight installation complete.

- Measure wires for trim motor and connect to motor
- Secure wires and cable inside fuselage to instrument panel

As you progress, you may have to revise your sequence plan. You may have left out some tasks, or you may have inadvertently put your cart and horse in the wrong order in your first draft of the plan. The important thing to remember is to look at system fabrication and installation one task at a time. One doable task doesn't look as formidable as the whole system. Even the tasks that you have laid out in your sequence plan do not have to be completed in one fell swoop. Set small goals for yourself, and if possible, try to accomplish something every day. Obviously, there will be days when you are unable to touch your project, but do your best to keep these days to a minimum. If you can do this, you will see progress, and you will have a sense of accomplishment that will keep you moving to the next goal, the next task, and the next, and the next. Now you are having fun (and we all know what happens when you are having fun). Before you know it, you'll have a system completed and be ready to start on the next system, task by task and goal by goal.

While working on your project, you are going to encounter numerous distractions. Some will be unavoidable. If you are one of the "great unwashed mass" and have to

**Here is what a finished MINI-MAX looks like.**

**A beautiful example of a Stitts Playmate, a modification of the Stitts Playboy.**

work for a living to support yourself, your family, and your project, you will probably be frequently distracted by having to go to your place of employment. Outside of winning the lottery or inheriting a fortune from Uncle Harry, you won't be able to do much about this. You will also have to devote a certain amount of time to your family, not to mention eating and sleeping in order to keep up your strength for working on your project. On the other hand, some distractions should be avoided if at all possible. For example, at one point in my construction process, I was working in my garage and the neighborhood kids were out of school for the summer. They would stop by to see the weird guy who was building an airplane. I showed them the plane, explained how it would really fly the broad blue skies one of these days, and answered their questions. It soon became obvious that the kids had a lot of time on their hands and were just stopping by to kill some of their spare time. I had to start working with the garage door closed in order to get something accomplished. Of course, it is important to interest the next generation in the aviation industry in general and the homebuilt movement in particular, but you don't want to overdo a good thing.

Adult friends can also be a problem. They are interested in what you are doing, and you certainly want to show off your skills as a builder, but again, don't overdo a good thing. Visits that become too frequent or too long can seriously detract from your building time. It is hard to tell friends to stay away, but at least try to keep working while you visit with them. If you can, give them simple tasks to do, even if it is just sweeping up or holding something for you. Watch out for the self-styled experts who know a better way to do something. If they are really experts in a certain area, take full advantage of their advice and suggestions. But if they are not experienced in the field they are expounding on, don't let them confuse you or talk you into something not specified by the designer.

A distraction that is insidious in nature but definitely controllable is daydreaming. A lot of time can drift by while you sit in your partially completed cockpit and dream of sailing over the hills and through the clouds on silver wings constructed with your own capable hands. Sitting on a stool at your workbench and admiring the fruits of your labor can also take up a disproportionate amount of valuable time. You do need a break now and then, and you do need to stop occasionally to consider where you are and where you are going on your project, but don't be caught in the daydreaming trap. Make your time productive—keep going and make every minute count! Make some progress every day, if you can, and one day you will come to the end of your sequence list. On that day, you will have a finished homebuilt aircraft sitting before you. Then you can take a minute or two to admire your outstanding handiwork and give yourself a well-deserved pat on the back.

Sitting in my half-finished Barracuda, dreaming of the wild blue yonder awaiting me and my bird.

Camping out with a Tailwind. You can't beat the price of the room, and your private air transportation is right at hand.

# Getting Information and Help

OW YOU ARE IN HIGH GEAR, going full speed ahead, and have the pedal to the metal on your dream plane's construction. But a homebuilt involves a wide variety of disciplines and techniques. Where do you turn when you have a problem? Whom do you ask for additional information? Where do you get the knowledge and skill you need to turn your project into a "living, breathing" aircraft?

The closest and best source of information and help is your local EAA chapter and the chapter's technical counselor. Not all EAA chapters have tech counselors, but most have one or more. The Tech Counselor Program was established by the EAA "to provide experienced individuals to advise amateur builders and restorers of aircraft in a technical advisory capacity, free of charge." Following World War II, when the Federal Aviation Regulations (FARs) were revised to include amateur-built aircraft, all in-progress (precover) inspections and final inspections of homebuilts were conducted by FAA inspectors. As the aviation industry grew by leaps and bounds, the FAA found that it had its hands full with inspections of the major airlines, commuter airlines,

corporate aircraft, flight training facilities, air freight lines, aircraft manufacturers, maintenance facilities, and air traffic control. At the same time, the EAA was growing by their own leaps and bounds, and the FAA was favorably impressed by the technical expertise displayed by many of the amateur builders.

In 1983, the FAA officially ceased precover inspections of homebuilts and published Advisory Circular (AC) 20-27, "Certification and Operation of Amateur-Built Aircraft" (see Appendix B). In AC 20-27D, the FAA states: "Amateur builders should have knowledgeable persons (i.e., FAA certified mechanics, EAA technical counselors, etc.) perform precover inspections and other inspections as

They say flying an open-cockpit biplane is an experience like no other. The wind in your hair and the sound of the wires makes for an unforgettable trip.

EAA chapters hold regular meetings and provide expert free advice.

appropriate. In addition, builders should document the construction using photographs taken at appropriate times prior to covering. The photographs should clearly show methods of construction and quality of workmanship. Such photographic records should be included with the builder's log and other construction records."

While the use of the word "should" throughout this paragraph makes it obvious that these actions are not mandatory, you may have a very hard time with your preflight FAA inspection if you have not complied to the best of your abilities with these recommendations. Make it easy on yourself and provide the best possible documentation of your building process (see Chapter 10).

The EAA has established certain qualifications for their tech counselors to ensure that they can provide the best assistance possible. A tech counselor must be a certified A&P (airframe and powerplant) mechanic or hold either the airframe or the powerplant certification alone. If not an A&P, the tech counselor must have completely built an amateur-built aircraft or completely restored an antique/classic aircraft. A tech counselor may also be an IA (an A&P with inspection authorization), a DAR (designated airworthiness representative of the FAA), a DER (designated engineering representative of the FAA), or an aerospace engineer. This is a voluntary, nonpaid program and the tech counselor does

**A Corben Baby Ace "A" model, powered by a 65-HP Franklin engine. The Baby Ace was popular in the 1950s and is still in demand today.**

not charge for visits or inspections. A builder may offer a cool or warm drink, depending on the weather, or offer to pay the counselor's expenses for gas, but no other compensation is allowed under this program.

The tech counselor will complete a report form (see page 40) and ask the builder to sign at the bottom. One copy goes to EAA headquarters and one copy is retained by the tech counselor. (The legal types have recommended that a copy of the report *not* be given to the builder for reasons of liability. Even though the tech counselor can only advise the builder about construction practices and techniques, no one wants the report to be used in a court of law in the event of an accident.) The builder should record in the logbook the date of the airframe inspection, the results, and the tech counselor's name, EAA number, and tech counselor number. Be sure that the visit and the results of the inspection are well documented. Keep these records in a safe and secure location so they will be available when time for your final inspection rolls around.

Few tech counselors are experts in all facets of amateur-built aircraft. Some have more experience in one or two primary construction materials and only limited experience with the other two. If the counselor is stuck for an answer, he or she should be able to help you find the answer. Remember, though, that the tech counselor acts only in an advisory capacity. The final responsibility for everything concerning a homebuilt plane lies with the builder and only the builder.

Your local EAA tech counselor will be your best source of help and information during your construction process and on into your flight testing, but the counselor will certainly not be your only source. Other members of your local EAA chapter may be more experienced in certain

A Lancair, one of the newer homebuilt designs that is built from a kit with many preformed composite parts. This one has a very complete instrument and radio panel.

## EAA TECHNICAL COUNSELOR REPORT

DATE _____ EAA TECHNICAL COUNSELOR _____

COUNSELOR _____ EAA # _____

Street _____

City/State/ZIP _____

Telephone _____

BUILDER _____ EAA # _____

Street _____

City/State/ZIP _____

Telephone _____

Aircraft Type _____

Aircraft Registration Number _____

Engine & Horsepower _____

COMMENTS _____

_____

_____

_____

_____

NOTE: The EAA Technical Counselor is a volunteer technical advisor only and in this capacity has no authority to approve or "sign off" the aircraft or any aspect of their construction.

The EAA Technical Counselor Program is intended solely to facilitate informal contacts between homebuilders of aircraft and interested persons who have informed the Experimental Aircraft Association and/or its Chapters that they are available to render advice to such homebuilders. The EAA Technical Counselor is not an employee nor agent of the Experimental Aircraft Association or its Chapters, and neither the Association or its Chapters make any representation as to the Counselor's experience or competence with respect to aircraft building or restoration.

**The builder agrees that he has read and understands the preceding, and consents to the distribution of this report to EAA Headquarters and any appropriate governmental agencies.**

_____
                    **Signature of Builder(s)**

Original — White for Counselor's records
1st Copy — Cream tagboard - postage paid return card to be mailed to EAA Headquarters.

**The EAA technical counselor report form.**

areas than the tech counselor. Don't pass up the chance to pick their brains. Have them out for a beer or a soda and ask their advice. You may even get a demonstration and some free labor out of them. Members skilled in certain areas often give demonstrations and lectures at chapter meetings. Don't miss these golden opportunities to add to your skill and knowledge. And don't be afraid to bring up questions at the chapter meetings. There is a good chance that someone may have an immediate answer for you or will be able to research it for you and get the answer back to you later. You may want to write or travel to another chapter for help. For instance, EAA Chapter 71, known as the Bakersfield Bunch in Bakersfield, California, probably has the greatest con-

centration of RV builders (RV-3s, RV-4s, and RV-6s from Van's Aircraft) anywhere in the world. If there is anything you want to know about building an RV, I'm sure someone there would have a ready answer for you.

On the other hand, don't overlook the designer and kit manufacturer. In most cases, they will answer your ques-

**This RV-4 shows its builder's skill.**

tions, since they have a vested interest in seeing that their planes are put together properly. Of course, it may take a little time. Be patient, and be sure to enclose a stamped, self-addressed envelope. They are busy people and may be getting a lot of inquires and questions.

There is also a vast store of knowledge about homebuilts at the world center of the homebuilt movement, the EAA headquarters. A letter to the editor of the *Technical*

**This all-metal RV-3, powered by an O-290G Lycoming engine, still needs a little paint.**

*Counselor News* should get you the information you need or at least point you in the right direction. Lists of articles, owners, and accident reports by aircraft type are also available from EAA Information Services.

In addition, a number of popular homebuilts have newsletters that are published by the designer, the kit manufacturer, or groups formed to support builders of that par-

The *EAA Technical Counselor News,* **a storehouse of technical data and tips.**

ticular design. If you are lucky enough to pick a design that has a newsletter available, be sure to take advantage of this valuable source of information. If no newsletter is available, find other people who are building or have built the same design on which you are working. If you contact a few, who knows, you may wind up publishing a newsletter for your group of builders.

Be sure to also read the periodicals written specifically for the homebuilt movement, such as *Sport Aviation,* the *Experimenter, Vintage Airplane, Warbirds, Sport Aerobatics* (all published by the EAA), *Kitplanes* (from Fancy Pub-

lications), and *Sportsman Pilot* (published quarterly by Jack Cox). They feature advertisements for innumerable books and videos for the amateur aircraft builder. Three books by Tony Bingelis—*Sportplane Builders, Firewall Forward,* and *Sportplane Construction Techniques*—are invaluable for the wealth of information they provide to the first-time builder.

In fact, there is so much good information available on homebuilts that you have to be careful not to become so ab-

**Some of the available periodicals that provide information on homebuilts, parts, supplies, and building techniques.**

sorbed in your reading that you neglect your building. You can check the front pages of this book for the address of the Iowa State University Press, which publishes Vaughan Askue's *Flight Testing Homebuilt Aircraft* as well as a number of Ron Fowler's books, all of which can improve your proficiency as a pilot.

I mentioned in an earlier chapter the valuable training and practice you can obtain from your local community college or technical school in such areas as welding, woodworking, riveting, sheet metal work, and working with composite materials. These courses take time but are well worth the effort and expense if you are not proficient in the skills required for your particular design.

The FAA is another source of important information, including Advisory Circular 20-27D, which I discussed earlier. And of course, the FAA is the only agency you can go

**Books, books, and more books are available to help you with your project.**

to when you are ready to start "the paper war" to get your dream plane certificated.

All in all, there are more sources of help and information out there than you can shake a control stick at. If you can think of the question, someone out there can think of the answers. After all, there is nothing new under the sun except that magnificent aircraft you are going to build.

One question you will undoubtedly have is "What engine [or engines] am I going to hang on this splendid airframe I have created?" In most cases, there are a number of possibilities and the final decision will be yours. In the next chapter, we'll look at the alternatives available to you.

A Q-200, powered by a 100-HP Continental O-200 engine. This engine can push this sleek little composite homebuilt along at a 180-MPH cruise speed.

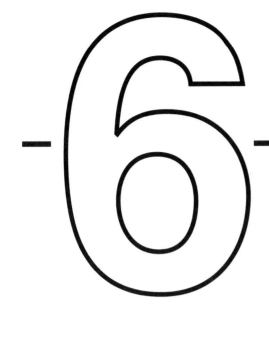

# Selecting an Engine

**I** **F YOU ARE DADDY WARBUCKS and money is not a problem, then selecting an engine for your dream machine isn't either. You just order the factory-new Lycoming or Continental of your choice, preferably the biggest one your plane will accommodate, and hang it on your engine mount. But if you are like most of us and you squeeze every penny until Lincoln hollers for mercy, you have to start looking for cheaper options for powering up that mean machine.**

If you are building an ultralight or a very light single-place plane, your engine selection will not put you into terminal sticker shock. You can purchase a Rotax, a small Continental, or a Volkswagen conversion without having to sell the house to pay for it.

When you start getting into engines of 100 HP and up, you need to shop around for the best deal that will suit your requirements. Most designs can accommodate a limited range of horsepower under the cowling. The Barracuda can fly with a 180-HP engine, and in South Africa, where homebuilts are limited to not more than 150 HP, one was built and flown with that size engine. (Every country has specific rules governing the construction of amateur-built aircraft, and some bar the practice completely. Appendix D provides a list of Canadian rules and regulations.) The Barracuda can also fly with engines up to 300 HP, and then it really hums right along.

On the other hand, the Long-EZ was designed specifi-

cally for the 115-HP Lycoming O-235 engine. Many builders initially install an engine on the low end of the allowable power range for reasons of economy and then, when they are disappointed with its performance, they save their pennies and switch to a larger engine as soon as possible. A good bet would be to go for the mid or high side of the allowable power range if you can afford it. It would be unwise to exceed the designer's specified range of engine power, either on the low side or the high side. An underpowered plane may not make it off the ground before the end of the runway jumps up and smites thee a mighty blow; and if you do manage to stagger into the air, you may not be able to fly out of ground effect without stalling. Definitely not a good position in which to place yourself.

On the other end of the scale, an overpowered plane can lead to severe weight and balance problems. You may have to add considerable weight to the tail to balance the big iron lump up front, and the empty weight of the finished

A nice installation of a Volkswagen conversion. You can see the lower cowling on the other side of the plane.

This 40-HP Rotax 447 engine fits nicely into the preformed fiberglass cowling on the MINI-MAX.

plane may approach or exceed the allowable gross weight. That means there is no room for you, the pilot, to go along with the plane. You may have built the world's largest remote control model plane, if that is the only legal way you can fly it. If you have 170 pounds (average pilot weight) left over to squeeze you in, there is still a good chance that the excess power will overload the structure beyond its ultimate design limits, resulting in catastrophic failure and your immediate demise. Another not-so-good position in which to place yourself.

Once you have decided on the size of the engine, there are a number of options open to you. Number one, Daddy Warbucks just pulls out his checkbook and orders a brand-spanking-new engine for $30,000 or more and waits for the truck to drop it off. The rest of us look at options two through five.

The second most costly option would be to buy a rebuilt engine with zero time since major overhaul (SMOH) and with all the accessories. These can be found at the factory, in your local engine shops, in the newspaper classi-

fieds, in *Trade-A-Plane,* or in aviation advertisements. There are a few things to watch out for here. What does "all accessories" include? What kind of guarantee do you get? What kind of reputation does the dealer have? This last one may be hard to pin down, but ask around. Bad news travels just as fast as good news.

The third most costly option involves finding an engine of the size and type you need that is used but still has time left until the next major overhaul is due. The cost will vary depending on the amount of time remaining on the engine. It may need a "top" overhaul, which consists of removing the cylinders and having them reworked by a reputable engine shop. This is much cheaper than a major overhaul. But why was the engine removed from its previous airframe? Was it in a wreck? Was it subjected to sudden stoppage with a prop strike? Is the crankshaft bent? Are the cylinders making good compression? Is the logbook in order? In other words, you had better have a powerplant mechanic check it out for you unless you have implicit trust in the person from whom you are buying the engine.

Moving right along to option number four, we come to run-out engines or engines that have flown off all their time to a major overhaul. If you can do the overhaul yourself or have a powerplant mechanic help you do the overhaul, you can save bunches of money. If you buy the engine and send it to an engine shop for the major, it is going to cost you bunches of money. As in option three, you need to know the history of the engine and you need to get a good, up-to-date

This Acroduster has a nice little 160-HP Lycoming O-320 engine installed. No word on the powerplant for the two-wheeler beside the tail.

logbook with it. The cost of the overhaul depends not only on labor but also on the parts that need to be replaced. If the engine is shot or the crankshaft is a throwaway, you are in store for some big bills. Again, as in option three, if you are not an engine mechanic yourself, you had better pay one to look at it before you plunk your money down.

Still too much money for you? Have I got a deal for

Here is a 200-HP Lycoming IO-360 that was run out to TBO and picked up for a song.

you! Go down to your local auto junkyard, rip out an engine for $500 to $1,000, and hang that puppy on your bird. Well, it's not quite that easy, but there are a lot of converted auto engines being flown these days as in the days of yore. Old Henry Ford kind of got things started when he began pumping out Model A and Model T Fords like hotcakes. Soon, amateur aircraft builders of the time were pulling the engines out of the cars and putting them into homebuilts like the Pietenpol Air Camper. The same thing is still going on today with Volkswagen, Corvair, Ford, Oldsmobile, Porsche, Buick, Mazda, Subaru, Honda, and Chevrolet engines. Have you flown a Ford lately?

There are advantages and disadvantages to using a converted auto engine in your flying machine. Taking the positive approach, let's look at the advantages first. The cost of the engine is going to be much, much less than an aircraft engine of comparable size. Overhauls will be about one-eighth the cost of the major overhaul on an aircraft engine. When you need a replacement part, you go to your friendly neighborhood auto-parts store and pay probably one-tenth the price that a comparable aircraft part would cost. Fuel consumption will also be considerably less for the auto engine, and you may be able to use the cheaper auto gas instead of av-gas. A note of caution: most auto gasolines now contain alcohol additives that will deteriorate the fuel lines, fiberglass fuel tanks, and seals used in many homebuilt fu-

**This Pietenpol Air Camper is propelled through the atmosphere by a Model A Ford engine.**

el systems. The Dow Chemical Company has a fiberglass resin, Derakane 470-36, which it says will remain untouched by alcohol additives (or oxygenated fuels), but you still have to worry about any rubber parts in your fuel system. In general, fiberglass tanks made with vinylester and polyester resins are resistant to deterioration from fuel additives, but epoxy resin is not. One more advantage, of an aesthetic nature, is that the Ford V-6 conversion is said to sound like the engine of a P-51 Mustang when it flies overhead.

On the downside, most of the auto conversions are liquid-cooled engines, and they therefore have the added complexity and weight of a radiator, cooling fluid, and associated cooling-system plumbing. When you weigh the auto conversion with all of its accessories, reduction drive, and cooling system filled with fluid, it is slightly heavier than a comparable air-cooled engine with its accessories. In addition, the size and shape of the auto engine may not be compatible with the cowling of your particular design and will require a special engine mount. The additional powerplant weight may require adjustment of the engine distance from the fire wall to keep the aircraft center of gravity within allowable limits, so this must be taken into consideration when the engine mount is being built. The auto engine will require some modification prior to installation. You will have single-point ignition giving you less redundancy than you have on an aircraft engine with two spark plugs in each cylinder. Most auto conversions must turn at high RPMs in order to produce the horsepower and torque required to get you in the air and keep you there. This mandates a reduction drive system on all but the smaller conversions, adding cost, weight, and complexity to the engine installation. Even with these disadvantages, more homebuilts with converted auto engines are taking to the wild blue each year.

What about a pusher-type engine instead of the usual tractor (puller) type? Some aircraft engines, such as the Lycoming O-290, can be used in the pusher or tractor position, but many engines can be used only in the puller position un-

**You can pick up a Ford V-6 engine with aluminum heads at your local junkyard. Note the belt reduction drive between the engine and the propeller.**

**This dynafocal engine mount is hanging on the fire wall of an RV-6 and waiting for a 200-HP Lycoming IO-360 engine to be bolted on.**

less they are properly modified. If you are going to build a pusher, be sure your propeller and your engine are compatible with that design.

**TABLE 6.1.  Some Engines To Choose From**

| Engine Manufacturer | Horsepower Range | Engine Manufacturer | Horsepower Range |
|---|---|---|---|
| Lycoming | 55-260 | Continental | 40-435 |
| Volkswagen | 35-50 | McCullough | 72 |
| Rotax | 35-100 | Subaru | 145 |
| Franklin | 49-240 | Evinrude | 85 |
| Warner | 125-175 | Hirth | 100 |
| Corvair | 145 | LeRhone | 80 |
| Pratt & Whitney | 220-2,500 | Ford | 60-200 |
| Fairchild | 200 | Wright | 150-180 |
| Outboard | 35 | Menasco | 125-150 |
| Siemans | 113 | Hispano | 180 |
| Buick | 220 | Chevrolet | 220 |
| Mazda | 180 | Oldsmobile | 220 |
| Walt Minor | 105 | Salmson | 40-135 |
| Dehaviland | 140 | Triumph | 40 |
| Anzani | 35 | Jacobs | 245-275 |
| Mercedes | 180 | Rolls-Royce | 100-1,180 |
| Renault | 230 | Heath AVN | 25 |
| Aeronca | 45 | Limbach | 68 |
| Gnome | 160 | Porsche | 75 |
| Micro Jet | 800 | Clerget | 130 |
| Lambeert | 90 | Leblond | 70 |
| Honda | 75 | Keickhafer | 80 |
| Ken Royce | 90 | Argus | 250 |
| Pollman | 40 | Nelson | 48 |
| Breda | 45 | Universal | 65 |
| Mercury | 70 | Cushman | 18 |

# 7

# Hitting the Wall

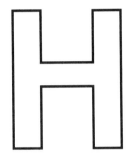

**AVE YOU EVER TRIED** any long-distance running? When long-distance runners have used up their first wind, they catch a second wind and keep on going. When the runners exhaust this second wind, it is said they have "hit the wall," and in many cases that ends the race for them. In some cases, though, the runner reaches way down to find some inner strength and keeps on driving ahead. Soon the wall is gone, and the runner has a third wind and probably a finish in the race. And even if this runner doesn't win the race, he or she doesn't quit.

Statistically, finishing a homebuilt aircraft isn't easy. It takes people who can reach down within themselves and find the inner strength to keep going when they "hit the wall"—not once, not twice, but probably many times when the tasks seem insurmountable. Some people who start homebuilt projects will never finish, never get to make that first flight in a plane they crafted with their own hands. There are a number of reasons for this grim reality: financial problems, deteriorating health, death (boy, that's a stopper), loss of interest, loss of medical clearance, other more pressing demands on your time, starting a project too ambitious for your particular skills, and the list goes on. But look on the bright side. Many *do* finish their projects. Many do make that first, exhilarating, adrenaline-pumping flight. Look at all of those beautiful homebuilts lined up at Oshkosh every year and at the regional and local fly-ins across the country. Someone just like you put those little

jewels together with their own hands. They did it, and you can do it too.

So, what do you do when you hit the wall? In Chapter 4, I discussed the way time tends to slip by and the need to keep a steady work schedule. Some of the same techniques that helped you keep up your steady progress can help you when progress grinds to a halt altogether.

Step number one: decide if you really want to finish the project. If circumstances have changed, or if you have personal problems that make completion impractical, you might as well bite the bullet and hang it up. You can advertise your project in the local papers or in the amateur builders magazines. You can't expect to get as much for all the bits and pieces as you have invested, to say nothing of a return on your time and labor. Count it instead as a learning experience. You may want to donate your project to a local EAA chapter and take the fair market value as a tax write-

A Stitts Playboy that is finished and ready to fly.

off (a charitable contribution to a nonprofit organization). Or, you can give the project to a friend and turn green with envy while you watch its completion. No matter how upset you may be, please don't chop up your project and throw it in the dumpster. Try to find a good home for it so it has a chance for eventual completion.

Now, if you have checked your innermost feelings, your yin and your yang, your pocketbook, your astrological chart and you have decided you really want to finish your unfinished masterpiece, you have to break through the wall. You have to get your third wind.

Step number two: dust off your project, pull out your plans, and look them over. Clean up your workshop and select a single, simple job that needs to be accomplished. Remember the long-distance runner. You've got to keep driving yourself and hang in there. If you don't, you're beaten before you start.

Step number three: get that simple job completed. Hey, that wasn't so bad. In fact, you enjoyed it. It was satisfying to see that simple task completed. Pick another simple task. Keep going, one simple job at a time. Remember the sequence of tasks you worked out in Chapter 4? If those planned sequences are still applicable, pull them out and follow them. If a job looks too formidable, break it down to simpler steps. Take it step by step, task by task, system by system, subassembly by subassembly, assembly by assembly.

Now the wall is coming down. You are starting to pick up momentum. It feels good again. You are looking forward

A prewelded fuselage structure for a Kitfox helps speed up the building process.

This Throp T-18 has been completed and finished off with a first-class paint job.

to getting to your workshop every chance you get. Your spouse is getting jealous of your project again. Things are back to normal. You made it through the wall!

But don't celebrate too soon. This may be only the first of many tests of your willpower and determination, so remember the three steps: 1) decide that you really want to proceed, 2) dust things off and get yourself organized, and 3) get busy on simple tasks and then just keep on trucking.

Maintaining motivation throughout the entire building process can prove difficult, but there are some techniques that can keep you pumped up and keep the wall from smacking you in the face. Take time to attend those monthly EAA meetings. The enthusiasm generated by a group of amateur aircraft builders is contagious, and you are bound to catch it. Also, as I said previously, this is the best place to get answers to your questions and to get any help you may need. Best of all, it motivates you to keep going.

Another highly motivating experience is a visit to air shows and fly-ins, especially those associated with home-builts and the EAA. Here you'll have a chance to talk to other builders—those who have beaten the wall and are piloting their magnificent creations to the fly-ins and those who, like you, are in the process of creation. You'll be able to see the finished products of each amateur builder's art, and you may find some examples of the same design you

are working on. In that case, you have really struck pay dirt. You can pick the brain of the builder and discuss problems and solutions. This can save you lots of time reinventing the wheel. You can benefit from someone else's mistakes and you won't have to make the same mistakes yourself. If possible, exchange names, addresses, and telephone numbers so that you can keep in touch. You may have a problem later that this successful builder can help you solve.

If you feel your motivation level slipping and there is no EAA meeting or fly-in coming up, try reviving your flag-

Mamma mia, that's a lot of joints to weld!

ging spirits by visiting another builder and talking shop for a while. You may even be able to help with something to get your creative juices flowing again.

Be careful that you don't spend so much time getting motivated that you don't have any time left for building. The motivation techniques are not an end in themselves but a means to an end—to keep you building.

Try to accomplish something on your project every day if you can. I've said this before, but it is important enough to repeat. The longer you stay away from your project, the harder it will be to get started again. Keep the momentum going and keep motivated. If you hit the wall, blast through that sucker and keep on going.

**At last, one wing is finished.**

# 8 Radios and Instruments

**W**HEN YOU START TO THINK about radios and instruments for your homebuilt, you are left pretty much on your own. Designers of plans and kits assume that the radios and instruments you install are an individual requirement, and they therefore provide little or no guidance.

Here is a simple instrument and radio arrangement in a single-place plane. Note the leather helmet over the control stick and the modern headset on the left.

Let's look at radios first. Here your choices range from no radio (zero, zip, *nada*) to a $100,000 stack of beautiful black boxes with lots of neat switches, lights, bells, and whistles. How do you determine what you need in your particular bird?

First, take a look at your project and the type of flying you will be doing. If you are building an ultralight or a very light plane so you can putt around the local area, you may not need any radio at all. In fact, a number of people have flown aircraft without radios all the way across the country without a great deal of difficulty. You just have to be selective about your route of flight and the airports at which you land. You don't want to stumble into a restricted area and have an F-15 come up to intercept you, nor do you want to try to land at Los Angeles International wagging your wings on final to indicate that you have no communication with the tower. Some homebuilts have no electrical system, so you are limited to battery-powered electrical marvels. Here is where a hand-held transceiver may come in handy. A hand-held radio takes up very little room, adds very little weight, and allows you to talk to someone on the ground or in the air if the need arises. Be sure that you get the 720-channel type. The old 360-channel radios are obsolete. You should also check with your airport manager at your home base about radio requirements. Even some small, uncontrolled airports frown upon non-radio-equipped aircraft flying in and out on a regular basis.

If your homebuilt carries more than one person and you intend to fly more than 50 miles from your home airport, you will need to install an emergency locator transmitter (ELT) as required by Federal Aviation regulations. If you are going to fly IFR (instrument flight rules), you will need a transponder. The FAA requirement on altitude reporting capabilities for transponders are presently in a state of flux, so before you buy one, check on the latest information from the FAA. If you plan on flying at or above 24,000 feet MSL (mean sea level) (some homebuilts can accomplish this feat), you will be required to install distance measuring equipment (DME).

For most of us, cost will be the determining factor in the quantity and quality of the radios we install in our majestic flying machines. According to the FARs, "[a] two-way radio communications system and navigational equip-

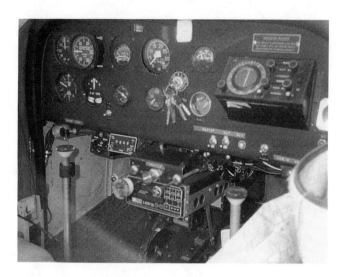

An older instrument and radio panel in an older Throp T-18.

ment appropriate to the ground facilities to be used" are required for IFR flight. A glance through *Trade-A-Plane* advertisements will show the vast array of communication and navigation radios available at a vast array of prices. The usual choice for navigation is a VOR (very high frequency omnidirectional range) with an ILS (instrument landing system) thrown in if you will be making instrument approaches to a landing. You will need an ADF (automatic di-

A modern instrument and radio panel in a modern Lancair. A set-up like this will cost some big bucks.

rection finder) if your flying is going to take you to (and from) a nondirectional beacon every now and then. Two newcomers to the navigation arena are LORAN (long-range navigation) and GPS (global positioning system). LORAN has been used for decades to guide ships around the oceans of the world, but only recently has the equipment been miniaturized to the point that it could be installed in aircraft. LORAN uses longitude and latitude coordinates to get you from one place to another. GPS uses signals bounced off U.S. Defense Department satellites to locate your position over the surface of the earth to within 60 feet or so. With either LORAN or GPS on board, you can throw the VOR and ADF away, unless you want them for backup in case of failure of the primary navigation system. Of course, if you are Daddy Warbucks, you can add other black boxes such as autopilot, weather radar, TACAN (tactical air navigation), RNAV (area navigation), MLS (microwave landing system), audio control panels, UHF (ultra high frequency radio), HF (high frequency radio), and a laser disk player with stereo headphones. Most homebuilts, however, can struggle along without these goodies and couldn't make it into the air at all with too many of them on board.

While you are making space in your instrument panel for your radios, you might as well consider putting in some instruments as well. After all, that's why they call it an instrument panel, right? The FAA in its infinite wisdom has specified the minimum instruments and equipment required for the various modes of flight, as shown in tables 8.1, 8.2, and 8.3.

### TABLE 8.1. Minimum Equipment for Flight under Visual Flight Rules (Day)

Airspeed indicator
Altimeter
Magnetic direction indicator (mag compass)
Tachometer for each engine
Oil pressure gauge for each engine
Temperature gauge for each liquid-cooled engine
Oil temperature gauge for each air-cooled engine
Manifold pressure gauge for each altitude-limited engine
Fuel gauge indicating fuel quantity in each tank
Landing-gear position indicator if aircraft has retractable gear

### TABLE 8.2. Minimum Equipment for Flight under Visual Flight Rules (Night)

All equipment required by Table 8.1 plus the following:
Approved position lights
Approved red or white anticollision light system
If operating for hire, at least one landing light
Adequate source of electrical energy for all installed electrical and radio equipment
Spare set of fuses, or three spare fuses of each kind needed

### TABLE 8.3. Minimum Equipment for Flight under Instrument Flight Rules

All equipment required by Table 8.2 plus the following:
Two-way radio communications system and navigational equipment appropriate to the ground facilities to be used
Generator or alternator of adequate capacity
Sensitive altimeter, adjustable for barometric pressure
Gyroscopic rate-of-turn indicator
Gyroscopic bank and pitch indicator (artificial horizon)
Gyroscopic direction indicator (directional gyro or equivalent)
Slip/skid indicator
Clock that indicates hours, minutes, and seconds with a sweep second-hand or digital presentation

The main instruments you are going to need can be separated into three groups: flight instruments, engine instruments, and navigation instruments. When talking about flight instruments, you often hear of "the basic six." These are the six instruments that show you what your plane is doing in reference to the rest of the universe. If you want to know how fast you are being propelled through the sea of air, you can refer to your airspeed indicator. If you want to know how high you are above mean sea level, you can check out your altimeter. To find out the direction you are heading, sneak a peek at your heading indicator. In case you are "slip-sliding around," your needle and ball will clue you in. If you have an attitude problem, check it out on your attitude indicator. Going up? Just ask your vertical speed indicator.

The standard military arrangement for the basic six instruments is shown on page 56. This arrangement is also usually followed in civil aircraft. But in homebuilts, hang on to your hat—anything goes! Many homebuilts do not have all of the basic six instruments installed. As you see in tables 8.1 and 8.2, airspeed and altimeter are the only instruments required for day/night VFR operations. If you are building an ultralight, you may not have anything except an airspeed indicator.

If you are installing the basic six, or most of them, it would be a good idea to follow the standard arrangement as much as possible. This makes it easier for a pilot switching from one plane to another to find various flight instruments. If they are out of order, your visual cross-check of your instruments will be disrupted until you get used to a nonstandard arrangement. Oh yes, don't forget a magnetic compass, which you should hang in the area of least magnetic interference.

Engine instruments will obviously depend on the type of engine you hang on your flying machine. Tables 8.1 and 8.2 show the minimum engine instruments for day/night VFR. Again, an ultralight doesn't have to meet these mini-

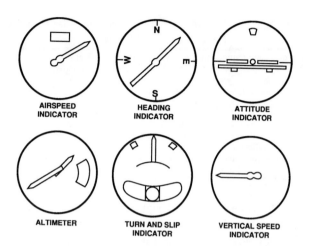

**Basic six instrument arrangement.**

AIRSPEED INDICATOR

HEADING INDICATOR

ATTITUDE INDICATOR

ALTIMETER

TURN AND SLIP INDICATOR

VERTICAL SPEED INDICATOR

mum requirements. Unless you are building a glider, you should install a fuel quantity gauge for each fuel tank. A simple sight gauge made from a piece of clear plastic tube will suffice in some cases. If your engine is a four-cycle type with oil separate from the fuel, you'll need an oil pressure gauge. If it is air-cooled, you'll need to keep track of the oil temperature as well to ensure against having an engine meltdown. For a liquid-cooled engine, you'll need a water temperature gauge, and a coolant-system pressure gauge is highly recommended but not required. A tachome-

ter will tell you how much power you are requesting from your engine. If your bird can climb to a rarefied altitude where you can overboost your engine, you will also need a manifold pressure gauge to enable you to restrict power to allowable limits at altitude. If you have a high-performance homebuilt, you may want to add an exhaust gas temperature (EGT) system on your hottest cylinder or on all cylinders. A cylinder-head temperature gauge can also give you a reading on how your hottest cylinder is doing. You may even want to install an engine analyzer system, which provides additional information on how each cylinder is operating.

Now, if you want to know where you are and where you are going, you will also need some navigation equipment and the associated gauges. An ADF set will have an ADF needle. A VOR, TACAN, or VORTAC will also have a pointer to show you the way. An ILS system will have a localizer needle and a glide-slope indicator combined into one gauge. LORAN and GPS will have indicators built into their control boxes. DME will require a gauge to tick off the miles to or from a selected location.

Now there are a few gauges that don't exactly come under the heading of flight, engine, or navigational instruments which you may or may not need. To monitor your electrical system, you will need an ampmeter and possibly a voltmeter. To track the flight time on your engine, you

**A nice, full panel on a Kitfox.**

**A full instrument and radio panel on a Cozy, a three-place composite homebuilt. Note the nicely upholstered seats.**

may want an hour meter (commonly known as a Hobbs meter). If your watch stops, a clock would be a handy item on your panel, and if you are going to get upside down and backwards (perform aerobatics), you'll need an accelerometer (g meter) to tell you when you are about to pull the wings off or just how hard that landing was.

While you are making room for all or some of these goodies, don't forget to leave room for the ignition switch, various and sundry electrical switches, circuit breakers, fuses, warning lights, panel lighting, engine controls, flap handle (don't put this in if you don't have any flaps), and a landing-gear handle (if you have retractable landing gear). You may have to put side boards on to hold all of these expensive items, but a better solution would be to keep your instruments and radios to a minimum and leave room to add extras later if you feel you need them. Remember, each black box or instrument adds complexity and weight as well as cost. Complexity means you may spend most of your time fixing things instead of flying your magnificent machine. Weight means your machine's performance will not be as magnificent as it would be if it were lighter.

If you have gyro instruments in your panel, they will have to be shock-mounted to protect their delicate inner workings. You can either shock-mount the entire instrument panel or you can shock-mount only the portion of the panel that supports the gyro instruments. The latter is usually the

**The backside of the Cozy panel shown above gives you some idea of the amount of wiring and plumbing it takes to set up a full instrument panel.**

**Back to basics for the pilot-builder of this Osprey amphibian. Note the paddle on the right. This keeps you from getting caught up a creek without one.**

**A simple panel in a single-place plane.**

easiest and best course of action.

Just in case (heaven forbid) something should go wrong with one of your instruments or radios or switches someday, you will need access to the back of your instrument panel so you can disconnect and remove the offending component. The easier it is to reach the back of your panel, the fewer bad words you will have to use while trying to get in there. You might consider a fold-down panel, a removable top cover, a removable gyro panel, or a combination of these to gain the necessary access. Standing on your head on the cockpit floor and reaching up into a mass of wires and tubing is less than satisfactory.

**A very full, very expensive panel with lots of gyro instruments.**

Before purchasing gyro instruments, you must decide on the system that will provide power to these instruments, vacuum or electric. Electrical instruments are great but very expensive. In either case, be sure the instruments you procure are compatible with the system that will make them run.

Sound complicated? Remember, all you have to do is take it one simple task at a time, step by step, system by system. Plan ahead and try to keep organized. It is easier to lay out your instrument panel on cardboard two or three times to ensure you have it right than it is to cut up a sheet-metal blank and then have to throw it away.

# 9 Finishing Touches

## T
HERE ARE OTHER THINGS besides the instrument panel layout that are usually left to the imagination and personal taste of the individual builder. These include seats, upholstery, access panels, cabin heat and vent, and the paint scheme.

Some kits include materials for pilot and passenger seats, but many do not. Seat construction will depend on the design of your homebuilt and will vary greatly. Some sim-

**A fairly simple seat designed for the Kitfox. Note the shoulder harness on the right seat.**

ple single-place planes get by with a piece of canvas stretched between a bar behind the pilot's shoulders to another bar under the knees. Some consist of a cushion on the seat pan and another behind the pilot's back. At the opposite end of the spectrum are fully adjustable seats taken from factory-built aircraft and installed in an amateur-built bird. One very important aspect concerning seats that is often overlooked in homebuilts as well as factory-built craft is crashworthiness. Pilots don't like to think about crashing their flying machines, but it is something that must be considered and planned for in advance of that terrible moment of truth. Planning for all possible emergencies may save your posterior. When building a seat structure for your plane, use all the latitude the basic airframe design allows to build in crashworthiness. In case you do contact the ground in an uncontrolled manner, you and your passengers will have a better chance of survival if the seats and airframe absorb most of the impact loads before these loads are transmitted to your frail and delicate bodies. These loads are absorbed by the compression of the seat and the airframe, so the more shock absorption you can build in, the better. Obviously, one thin cushion isn't going to do much, and your body weight will compress even a thick piece of foam rubber until there isn't much between you and the seat pan. Think along the lines of springs or a frame that will compress under impact loads but hold up under your weight (plus four or five g's if you are doing aerobatics or don't always grease your landings).

These plush seats are in a Lancair and come as part of the kit.

A restraint system that will keep you secured in the seat is also a must. Make sure your seat belt is secured to hard points in the airframe structure that are not likely to rip loose under crash loads. All restraint systems should include a shoulder harness with an inertia reel that will automatically lock under high transverse g loadings (a sudden stop that will try to imprint your face on the instrument panel). The shoulder harness should be secured to a hard point behind the seat and level with or above the occupant's shoulders. Some designs do not incorporate provisions for a shoulder harness, but it is a vital safety consideration and should be added as a modification if necessary.

You should also consider crashworthiness when you

These seats are in an Osprey amphibian. Note the solid-looking seat belt-shoulder harness arrangement.

are installing your instrument panel, flight controls, and engine controls in the cockpit. If at all possible, you should not have sharp edges or protruding controls or handles that could injure you or your passengers in case of a crash landing. Try to make everything in the cockpit as flat, as soft, and as smooth as possible in case you come in contact with it in a less-then-controlled landing.

And while you are busy color-coordinating your cockpit upholstery with your crashworthy seats, give a thought to soundproofing your office in the sky. Some lightweight insulating material behind the upholstery and on the aft side of the fire wall will help not only to reduce engine and wind noise but also to slow the transfer of heat or cold from the outside world into your cozy environment. Make sure, if you can, that the insulation is fireproof or fire retardant and does not give off toxic fumes when exposed to an open flame. A good seal around the canopy will also reduce wind noise when you are skimming through the sky. Bill Buethe, who sells plans for the Barracuda, devised a canopy seal made from rubber tubing and a squeeze ball from a blood-pressure testing belt. After he closes and locks the canopy, he pumps up the tube to provide a tight seal. If you have an aluminum skin on your homebuilt, you might consider gluing light rubber panels on the inside of the fuselage to prevent noisy oilcanning.

To ensure that your cockpit is cozy under most conditions, you will want to provide some means of cooling and heating the interior space. Cooling can be accomplished by venting outside air into the cockpit. This is usually incorporated into the basic design, but if not, you will have to add this necessity yourself. Check the placement of vents on other homebuilts of the same or similar designs to determine the most-effective locations and methods of installation.

Cockpit heating is usually provided by installing a heat muff around some portion of the hot exhaust pipes and ducting this heated air into the cockpit to keep you warm and toasty. This system requires frequent and careful inspection and maintenance to ensure against an exhaust leak that would allow carbon monoxide to enter the cockpit along with the warm air from the heat muff. Carbon monoxide is an odorless, colorless gas that will put you into a sleep that you won't wake up from until your next life as a mill worm or something.

Your basic design will probably include some access panels in the wings, tail, and fuselage so that certain critical areas can be inspected periodically and repaired if neces-sary. Look at the number and placement of these access panels with a critical eye. As the builder of this mechanical marvel, you will be qualified to inspect and maintain it for as long as you own it (see Chapter 10). Imagine having to look through these access panels with a bright light and an inspection mirror. Imagine having to reach through these access panels with tools and a work light to remove a cotter pin, take out a bolt, remove a worn rod-end bearing, and in-stall a new one. Could you do it? Is the access panel big enough? Is it in the best location? Consider the possibility of increasing the number of access panels and enlarging or relocating them. No modification to the design of the air-craft as envisioned and engineered by the designer should be undertaken lightly, but as the builder, you have the pre-rogative to make any changes you may wish. Only the FAA inspector performing your before-flight inspection can tell you that you can't if the modification makes your plane un-airworthy. In most cases, your modification will be hidden away by then and will in no way hinder you from charging off into the air—and killing yourself if you made a gross er-ror with your change. Therefore, unless you are an expert in the particular area that you plan on changing (structural, aerodynamic, controls, electrical, hydraulic, or what have you), get advice from someone whose knowledge and ex-pertise you trust. This may be the designer, your EAA tech-nical counselor, an A&P mechanic, or a professional engi-

You won't have any prob-lem getting to the engine or the exhaust pipes on this French designed Cri-Cri.

**Jimini probably rides inside for cross-country flights.**

**This great-looking Cassutt Racer has been modified to look like a P-51 Mustang.**

neer. Certain simple changes can be made on your own, but if they affect the structure, aerodynamics, or flight controls, proceed with utmost caution and get all the expert advice you can.

Even designers sometimes take a cavalier attitude toward the manner in which they make a modification to their prototypes. One designer-builder of a small, all-metal, open cockpit homebuilt (which shall remain nameless to protect the guilty) thought his head was sticking too far up into the slipstream. Before his next flight, he took a pair of tin snips

to the sheet-metal seat pan and cut out a large hole. He then bent the sharp edges down with a mallet, put a thin pillow into the hole, put his posterior into the hole, and took a hop around the pattern again. On landing, he proclaimed the modification to be a success. His head was out of the wind and he didn't fall out through the bottom of the fuselage.

The last finishing touch required on your creation will be the paint job. Here is your chance to really shine, so to speak, with a spectacular finish. First, if you are painting wood, fabric, metal, or fiberglass, any imperfection in the

**This bright red Falco has been finished with a beautiful paint job.**

**RVs are often modified to fit the builder's desires, but you should check with the designer before making major modifications. From top to bottom, an RV-3, an RV-4, an RV-6, and an RV-6A.** Photograph courtesy of Van's Aircraft, Inc.

**The large stripe on this Pulsar makes it look like the go-fast machine that it is.**

surface will show through the paint. Therefore, most of your time will be spent in preparing the surface for paint. Don't load your poor bird up with filler to smooth out the surface or it may be too heavy to waddle off the ground. If you decide to paint it yourself, as opposed to having a professional do it, you can build an inexpensive spray booth out of a two-by-two wooden frame covered with clear plastic. You should rent or buy the best spray gun that you can afford. It really makes a big difference. Practice a little on some similar material until you have a feel for the operation and the adjustments on the gun. Auto paints are much cheaper than aircraft paints and seem to give just as good results. Follow the instructions for priming coats and the mixing of paints carefully. Most long-lasting paint jobs will require a two-part epoxy paint, and if you breathe too much of the fumes from these epoxies, you can cause permanent damage to your lungs. Get the type of respirator mask recommended for the paint you are using and don't scrimp on your costs here. It will be cheaper than a double lung transplant. You'll need an exhaust fan and a high-volume compressor for your paint booth. Paint fumes can be flammable, so be sure the compressor and the motor on the exhaust fan are completely isolated from the fumes. You don't want to turn yourself and your plane into crispy critters with a flash fire caused by a spark from your compressor motor or your fan motor. Spray painting is an art when done correctly, so if you don't feel comfortable in this area, leave it to a professional.

You think you are ready to fly when the paint dries? Not so fast! Remember the cartoon of the little kid on the potty? The job's not finished until the paperwork is done. Next we'll fight the paper war.

# 10 The Paper War

BOUT SIX MONTHS before the scheduled completion of your plane, you should start the paper war in earnest. Actually, you started long ago when you first purchased your plans or your kit. That is when you started a file of receipts for your kit, your materials, your parts, and your supplies. You did save all of your receipts, didn't you? That's part of your record of construction that the FAA representative will want to see during your final before-flight inspection. You will also need the file of photographs you have taken all through the construction process.

It is recommended that you keep a construction journal in a three-ring binder showing the work you performed. Here's a typical entry: February 8, 1994, spent two hours building ribs. A book such as this which includes your receipts and photographs will provide the necessary proof that you have complied with the 51 percent rule—that you actually built at least 51 percent of the aircraft and therefore are eligible to request an airworthiness certificate under the experimental aircraft category. You will be required to sign a notarized statement to this affect (FAA Form 8130-12, Eligibility Statement—Amateur-Built Aircraft) as part of your application for an airworthiness certificate, so don't cheat.

You could be subject to a $10,000 fine and five years in a jail cell if you are caught making a fraudulent statement.

You will need a logbook for your aircraft, your engine, and your propeller and accessories (assuming you are not building a glider). You can pay a bunch for nice leather-bound logbooks or you can spend a little for paper-bound books. They are available from any pilot supply shop, most homebuilt-parts supply houses, or EAA headquarters. Be sure to make an entry in your aircraft logbook for each inspection you have during the construction period. If your EAA technical counselor gave you anything in writing, attach this to your logbook; if not, just make a pen-and-ink

**Take plenty of pictures showing your construction skill as you go along.**

entry showing the date, the inspector's name and tech counselor number, the results of the inspection, and your signature. Be sure to make a note of all approvals to close any area that cannot be inspected after being closed or covered.

**How many wings did you say this plane has?**

Advisory Circular (AC) 20-27D, Certification and Operation of Amateur-Built Aircraft (see Appendix B), in paragraph 11C states that the builder must "substantiate that the construction has been accomplished in accordance with acceptable workmanship methods, techniques, and practices. It is recommended that this evidence be documented in some form (e.g., the *Service and Maintenance Manual* available from EAA)." Table 10.1 shows the contents of the *EAA Custom Built Aircraft Service and Maintenance Manual.*

As you can see from Table 10.1, the *EAA Service and Maintenance Manual* documents what went into your homebuilt in the way of materials, parts, and supplies. This information will be required for you to properly service and maintain your plane over the ensuing years of flying. The FAA representative will want to see that you have kept a complete record of all the parts, components, fuels, oils, and lubricants that have gone into your flying machine. Some greases turn to concrete when mixed, so you have to know what was used originally to keep from using the wrong grease, oil, brake pad, or whatever later on.

**TABLE 10.1.   Contents of the *EAA Custom-Built Aircraft Service and Maintenance Manual***

General Description of Aircraft
General Builders Data—Sources of Materials
Specifications
   Design Data
   Performance Data
   Equipment List
   Weight and Balance`
   Rigging
Float Plane or Amphibian Data
Maintenance and Operation
   Lubrication, Landing Gear
   Brake System
   Electrical System
   Controls and Instruments
   Fuel System, Accessory Items, Fuel System Schematic
Repairs
   Repairs Log
Engine Troubles and Their Remedies
EAA Safety Checklist

At some point prior to your final inspection, you will have to conduct a fuel-flow test and document the results to show that you will have sufficient fuel flow to the engine at full throttle, at a steep angle of attack, and with a low fuel quantity in your fuel tanks. Details of how to conduct the test are clearly presented in Tony Bingelis's book *Firewall Forward* and in Vaughan Askue's *Flight Testing Homebuilt Aircraft.* The latter also tells you how to calibrate your fuel system, check for leaks in your pitot system, rig your flight controls, calculate weight and balance, and align your landing gear prior to your final inspection. The FAA representative will want to see the results of your fuel-flow test and your weight and balance calculations. Weight and balance calculations will show that your homebuilt will not exceed the design gross weight limits or the CG limits under normal operating conditions. Forms for these calculations are included in the *EAA Service and Maintenance Manual* and in AC 91-23A, *Pilot's Weight and Balance Handbook,* available from the FAA.

**The tail of this RV-6 makes a handy table. Very nice flush rivet work on this bird.**

A fireproof identification plate must be attached securely to the outside of the aircraft so that it will not be defaced, removed, or lost during normal service or as the result of an accident. The plate must be located adjacent to and aft of the rear-most entrance or near the tail surfaces and must be legible to someone standing on the ground. The following information must be etched, stamped, engraved, or marked in some other fireproof method: (1) name of the amateur builder, (2) model designation, and (3) serial number selected by the builder. My ID plate will read

Don Walter
Barracuda
S/N 504

The serial number was selected to correspond to the number on my set of plans from which the Barracuda is being constructed. Your serial number can be any number you wish to use.

A passenger warning placard must be mounted in the cockpit (except for single-place planes) so as to be in full view of all passengers. The stirring words, obviously designed to instill total confidence in your passengers, are as follows:

PASSENGER WARNING—THIS AIRCRAFT IS AMATEUR-BUILT AND DOES NOT COMPLY WITH FEDERAL SAFETY REGULATIONS FOR STANDARD AIRCRAFT

As if this isn't enough to scare off even the bravest soul, you must further inform your erstwhile passengers that they are about to embark in an experimental aircraft. You must also place the word EXPERIMENTAL in letters at least two inches high (and not to exceed six inches) at each and every access to the cockpit area. The ID plate, passenger placard, and EXPERIMENTAL decals can be ob-

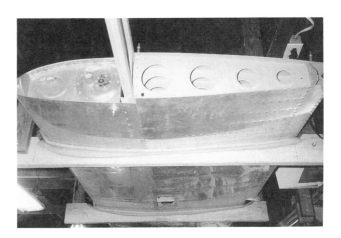

**Wings usually get stored in the rafters until it is time to install them. Note the curved and padded supports that protect the aluminum surface.**

tained from EAA headquarters or from most amateur airplane-builder supply houses.

Now that we've scared away all possible passengers, we can take a look at our instrument panel and make sure we have at least the minimum instrumentation discussed in Chapter 8. The instruments must be properly marked to show the operational limitations that apply to your particular aircraft and engine. These are such things as the so-called never-exceed speed, maximum flap-lowering speed (if you have flaps), normal oil pressure range, maximum oil temperature, and so on. If any instrument markings are on the glass front of the instrument, a white slippage mark must be placed on the instrument overlapping the glass and the instrument case. This will show you if the glass has slipped in relation to the case, causing the markings to shift from their normal positions.

All engine controls must be marked to show the proper direction of movement, such as *full* and *off* for the throttle and *full rich* and *idle cutoff* for the mixture. Moreover, they must move in the standard direction, such as forward

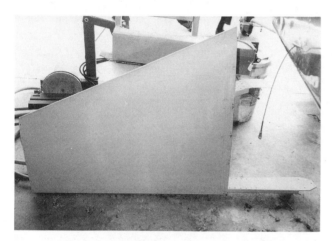

This vertical stabilizer has been covered with its aluminum skin and waits patiently to be finished.

for *on* and aft for *off*. Landing-gear handles, flap switches, ignition switches, master switches, and carburetor heat must all be marked for the proper operation.

An ELT, with a current battery, must be installed (except for single-place aircraft) as explained in Chapter 8, and the battery's expiration date must be marked on the battery. A cockpit checklist must be available in the cockpit to specify normal and emergency procedures and any operational limitations that are not shown by the instrument panel markings.

Now, think back to that benchmark moment about six months prior to your scheduled completion date. That is the time you should have rushed down to your local FAA office and picked up a copy of Aeronautical Center (AC) Form 8050-88, Affidavit of Ownership for Amateur-Built Aircraft, which is really a request for a registration number for your little jewel. If you want a special number, you'll have to submit a letter listing five possible numbers you would settle for and enclose a fee of 10 dollars. It will then cost you 10 dollars each year to hang on to your special registration number. If you are building from a kit, you should enclose a copy of your signed bill of sale for the kit, or else

you will have to fill out AC Form 8050-2 in place of the bill of sale.

Now, you have the ball rolling. If someone doesn't drop the ball, you will eventually get your requested registration number in the mail along with (what else?) another form to fill out, AC 8050-1, Aircraft Registration Application, and some other information requests. Fill out all the paperwork, keep the pink copy of AC 8050-1, and send the rest back with a five-dollar fee (it may be more by the time you read this, certainly not less). The pink copy is your temporary aircraft registration and is good for 90 days. If you don't receive your AC Form 8050-3, Certificate of Aircraft Registration, by the time 90 days have elapsed, give the FAA a holler and find out who dropped what through which crack in whose floor.

You can put your registration number on your plane as soon as you receive it. Be sure to follow FAR 45.22 and 45.25 for fixed-wing aircraft and FAR 45.27 for nonfixed-wingers as to the size and location of the numbers. Basically, they can go on the vertical stabilizer or the side of the fuselage and can be three inches high if your maximum cruise speed is less than 207 MPH (180 kts) and 12 inches high if your little rocket exceeds this speed criterion.

One more thing, if your homebuilt has a radio, you will need to contact the Federal Communications Commission to obtain a radio station license for your aircraft. Pilots no longer need a radio operator's license as they did in the past.

Now you are ready to apply for your final inspection and your airworthiness certificate. And what do you need? All together now, "Some more forms!" Fill out FAA 8130-6, Application for Airworthiness Certificate, and FAA Form 8130-12, Eligibility Statement—Amateur-Built Aircraft. This last one has to be notarized because you are swearing on a stack of logbooks that you built at least 51 percent of the flying machine. Enclose a three-view drawing or a photograph to identify the aircraft to be inspected as well as a letter stating that you have everything ready for the inspection and identifying the area where you intend to perform your flight test. The flight-test area should be within a 25-statute-mile radius of the airport you will be using, over an unpopulated area, and away from airways. Samples of these forms and letters are shown in Appendix B, Advisory Circular 20-27D. When you pick up your forms from the FAA, check to see if AC 20-27D is still current. If not, get the latest revision before you start the paper war.

While you are at it, make sure AC 65-23A, Certification of Repairmen (Experimental Aircraft Builders), is also

This Woody Pusher has won the paper war and is now enjoying the fly-in circuit every summer.

current (see Appendix C) and apply for that certificate. With this in your pocket, you can legally maintain your homebuilt, perform the annual inspection on the airframe and engine, make repairs, and make modifications with the same care and caution previously discussed. This privilege extends only to the plane you have built, so don't get carried away and start signing off inspections for other people unless you are a duly certified A&P or IA (inspection authorization). And be sure you document everything you do to your plane in your logbooks.

Before your FAA final inspection, get a final inspection from your EAA technical counselor, A&P mechanic, or

A Pietenpol soaks up some warm California sunshine at an EAA fly-in.

whoever has been doing your before-closing inspections. Have the bird opened up just as you will for the FAA representative and ask your inspector to go over your creation with the proverbial fine-toothed comb. Don't get your tail in a kink if you receive a long list of things that need to be corrected. That's what your advisor is there for. You want everything to be as good as you can make it before the FAA final, so you have to do it only once. If you have a designated airworthiness representative (DAR) for your final inspection, which you probably will with the present shortage of FAA personnel, the inspection will cost you some bucks. If the DAR has to come back a second or a third time, it will cost you some more bucks for each visit until you get it right. Try to get it right the first time out of the chute. Have all your paperwork in order, have your bird ready, and who knows, you may win the paper war without any casualties.

#### TABLE 10.2.   Checklist for FAA Inspection

1. Receipt file for materials, parts, and supplies
2. Photo file of construction process
3. Construction log
4. Logbooks: aircraft, engine, and propeller
5. Logbook entries for inspections during construction
6. Service and Maintenance Manual
7. Record of fuel-flow test
8. Completed weight and balance (part of item six)
9. ID plate on aircraft
10. Passenger warning placard (except single-place aircraft)
11. Experimental placards on each cockpit access
12. Minimum instruments required
13. Proper instrument markings for operating limitations
14. Proper control markings
15. ELT (except single-place aircraft)
16. Cockpit checklist
17. Registration number on aircraft
18. Registration (pink copy of request or AC 8050-3)
19. Radio station license (if radio equipped)
20. FAA Form 8130-6, Application for Airworthiness Certificate
21. Photographs or three-view drawings to identify aircraft
22. Notarized FAA Form 8130-12, Eligibility Statement, Amateur-Built Aircraft
23. Letter identifying the aircraft and specifying the test area

When the FAA representative is satisfied with your paperwork and the airworthiness of your aircraft, you will receive two sets of operating limitations, one for the test period (phase 1) and one for operations after the test period (phase 2). Builders are allowed to complete the test period and then proceed to phase 2 (normal operations) based on the assumption that they will honestly comply with all requirements of phase one.

Now comes the fun part. You get to fly this miracle machine you have built with your own hands, with tender loving care, with blood, sweat, and tears. Try to remember, did you make all those glue joints good and strong? Are all the shop heads on your rivets the proper height and diameter? Was that fiberglass layup wet enough? How was the penetration on those welded joints? Well, you are soon going to find out, because you are going to test-fly this baby!

**A beautifully completed Pulsar shows its stuff in the air.**

# 11 Testing Your Creation

**T**HE FAA REQUIRES that an amateur-built aircraft be test-flown for 25 hours if it is powered by a certified aircraft engine-propeller combination and for 40 hours if it has an uncertified engine such as a converted automobile engine. If you have built a glider, balloon, dirigible (a dirigible?), or an ultralight; you will be required to complete a 10-hour test period and at least 10 takeoffs and landings.

During the test period (phase 1), you are not allowed to carry passengers and you must fly only in the test area specified in your special airworthiness certificate. You must log the flights and maneuvers accomplished during these flights so that you have a record of having tested your aircraft under all of the conditions you expect to encounter in normal operations (phase 2). In other words, if you want to fly at night, not only must your bird be properly equipped for night flying but you must demonstrate during the test that it is capable of being flown safely at night. If you want to do spins or aerobatics, you must perform during phase 1 all the types of spins—upright, inverted (gulp), flat (double gulp)—and basic aerobatic maneuvers (rolls and loops) that you plan to perform after phase one. The basic aerobatic

maneuvers performed during the test period can be combined into more complex maneuvers later. Of course you can have your permanent airworthiness certificate modified at a later date, but it is easier to get everything included during your test period.

There are some basic preparations you need to take care of before your first taxi test. If you haven't been flying on a regular basis during your construction phase, you will need to brush up on your flight proficiency, ensure that your biennial flight review is current, and make sure your medical certificate is also current. If you have built a taildragger, you should do your brushup in a taildragger. They are definitely temperamental when it comes to ground handling, and some are more squirrelly than others.

**This Sidewinder has finished its test program with flying colors and has also been painted with bright flying colors.**

Review the normal and emergency procedures included in your operating manual. Either buy or rent a parachute and helmet and practice bailout procedures. I know, I know, you don't plan to make a nylon letdown after all the time and money you put into your flying machine, but you need to be ready for any possibility. Besides, you need to be around to collect the airframe insurance. You did insure it didn't you? How are your PLFs (parachute landing falls)?

Any skydiver or military jock can demonstrate a PLF to you in about five minutes, and it keeps you from breaking a leg or an ankle on landing. It's easier to walk to the nearest telephone if nothing is broken.

When you select the airport you will be using during your test period, check on available emergency equipment. If it is a small airport with little or no emergency equipment, you should have a friend standing by with a good-sized fire extinguisher and a list of emergency numbers. If something does go wrong, it is not the time to be paging through the telephone book, and you may not have the emergency 911 service in the area. Don't forget to stuff some change in your buddy's pocket if the phone is a pay phone. Keep the number of friends to a minimum during your ground and flight tests. You don't want a big crowd waiting for you to perform superhuman feats of daring. The peer pressure could push you into doing something you shouldn't. You should get approval from the airport manager to conduct your special tests, and you will probably need to do your tests when the airport traffic is at a minimum. Be sure you tell the tower (if there is one) or announce on unicom exactly what you are doing. You don't want any surprises for yourself or for anyone else.

One way to eliminate surprises is to do a very thorough preflight before every ground test or flight test and a very thorough postflight afterward. Make sure everything is right

**A Long-EZ says a little prayer that the clouds won't rain on its open cockpit.**

before each test, and then check to see if anything went wrong during the test. Naturally, if something is amiss, correct it before continuing. Don't press. If you don't feel right, if the bird doesn't feel right, if the wind or weather conditions are not right, or if the sun is sinking in the west, don't press your luck. Wait for another day, if necessary, when everything is in your favor.

In fact, you may not feel up to flying off the test time on your homebuilt at all. There is no dishonor involved if you have a more experienced or a more proficient pilot perform the tests or some of the tests for you. You will, in fact, be a test pilot when you perform the first taxi test, the first flight test, the first stall, or the first anything, and this may not be your forte. Unexpected things happen, and if you feel you are not fully capable of coping with the unexpected, find someone to iron out the rough spots for you. There will be plenty of time for you to get your stick time. On the other hand, if you take the tests slowly and step-by-step, an average pilot should be able to handle anything that comes along. Just don't exceed your limitations or the limitations of your aircraft.

You can divide your test program into three parts: taxi tests, first flight, and expansion of the flight envelope. Taxi tests are divided into slow taxi, fast taxi, and land-backs. Before you start on your slow taxi test, select an abort point that will allow you plenty of room to stop from the 40-knot airspeed you are going to try to achieve. Check your brakes as you start to roll out of your parking spot, and make sure your engine checks out in the run-up area. Before you are cleared on the runway, announce your intentions to the tower or on unicom. You are going to tie up the runway longer than a normal takeoff would, so don't surprise anyone who is trying to land. Use 1,500 RPM on the throttle for the first run and release the brakes. If you encounter any problem, reduce power to idle at once and clear the runway if possible. If you don't get to 40 knots by the time you reach your abort point, chop the throttle, clear the runway, and find out why you are not accelerating properly. If everything works as it should and you reach 40 knots, reduce your power, slow down, and taxi back for another run. On each successive run, increase the power at the beginning of the run by 300 RPM until you are going to full power, accelerating to 40 knots, and then coming to a stop without any problems. If you are in a taildragger, use these runs to practice keeping that nose straight down the runway.

Now you are ready to graduate to fast taxi. Using full power, accelerate to 40 knots as before, and reduce power

**A magnificent Glassair III waits for its next chance to dance with the clouds.**

to idle as before, but now, lift the tailwheel off the ground (or take the weight off the nosewheel if you are in a nose-dragger). Then let the tailwheel or nosewheel settle back on the runway and brake to taxi speed to clear the runway. On the next run, reduce the power only halfway before you lift the tailwheel or nosewheel. Then bring the power to idle and taxi back. On the third run, leave the power at full throttle as you lift the tail or nose, and then reduce the power to idle and taxi back. Starting to get dizzy from going in circles? Take your time. Don't rush it, and above all, if you encounter anything out of the ordinary, fix it before proceeding further.

Now you are going to push the speed above 40 knots in 10-knot increments until you reach 1.3 times your estimated stall speed ($V_s$). This usually works out to be about 60 knots for most homebuilts. Once you have accelerated to 1.3 $V_s$ with the main gear still firmly planted on the runway, you are ready to try some land-backs. You are going to finally get your plane in the air, even if it is only for a few seconds. As before, accelerate to 1.3 $V_s$ with the tailwheel up or the weight off the nosewheel. Then reduce power to idle, gently lift your bird off the runway, and gently allow it to settle back to terra firma. On the next run, leave on a little more power so that you stay in the air a little longer. Don't push too close to the end of the runway on the rollout. During the few seconds you are airborne, you should check for any yaw problems or a heavy wing. Any control problems will, obviously, have to be corrected before further testing. Finally, maintain full power while staying about three feet above the runway to a predetermined point that will allow you sufficient room to reduce power to idle, touch down, and brake to taxi speed before reaching the end of the runway.

Now, after you have corrected any and all engine and airframe problems you encountered during the taxi tests, you are ready for that one giant step for all aircraft builders—The first flight! Author Vaughan Askue in his excellent book, *Flight Testing Homebuilt Aircraft,* recommends that you prepare a flight card for each test flight, and I wholeheartedly agree. The flight card should list step-by-step what you wish to accomplish on the flight with the airspeeds and altitudes you are looking for along the way.

Start with a careful check on yourself, your equipment, the weather, and your aircraft. Are you feeling in tip-top shape? If not, wait until you do. Is all of your equipment ready? If not, correct the deficiency before proceeding. Does the weather meet your preset criteria? If not, wait until it improves. Did your meticulous preflight turn up any problem? If it did, fix it and document the fix in your logbook. Review in your mind the possible emergency landing sites near the airport, especially right off the end of the runway. Select a climb and power-off glide speed equal to 1.4 $V_s$ and call it $V_O$. Be sure your friend with the emergency numbers and pocket full of change (and the car keys) is fully briefed, and be sure the airport authorities are aware of and have approved your special activities.

By George, I think you are ready to do it! Dry your sweaty palms on your flight suit and fire that puppy up. After your run-up, do one more full-power land-back to get the feel of things, and then you are ready for the real thing. Usually, the first flight should include these and only these maneuvers: 30-degree- and 45-degree-bank turns at your $V_O$ (1.4 $V_s$) speed, and 30-degree-bank turns at 1.3 $V_s$, gentle stalls (level, in turns, and with landing flaps), one go-

This Moni powered glider doesn't need a tow plane to get it airborne. It does it all on its own.

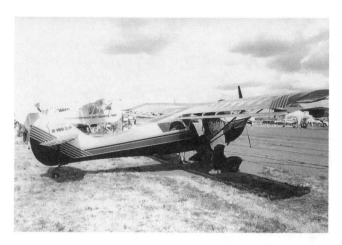

The Kitfox gets off the ground in a hurry with the help of the high-lift devices on the wing.

around, and a full-stop landing. Of course, if any problem develops, land as soon as practical and correct the problem. If the landing gear is retractable, leave it down on the first flight. If the flaps give you problems, land without flaps. If you can get approval from the tower, do your air work at 4,000 feet above the airport. If not, stay as close to the approach end of the runway as possible over an unpopulated section of your test area.

So, here we go! Apply full power, take off, climb straight ahead to 1,000 feet AGL (above ground level) at $V_O$ speed, and perform a nice, coordinated, 30-degree-bank turn to downwind. Check your instruments for problems, especially engine overheating. If there is something wrong, you are in a perfect position for landing. If everything looks fine, exit the pattern, climb to 4,000 feet AGL (preferably over the airport), and level off while maintaining $V_O$. Try some 30-degree-bank turns and then 45-degree banks. Slow to 1.3 $V_s$ and try 30-degree-bank turns to simulate your landing pattern. Still at 1.3 $V_s$, try a couple of stalls straight ahead to see what will happen if you get too slow on final. Check the altitude you will lose in the maneuver. Now accelerate to $V_O$ and lower half-flaps, and go through the 30-degree- and 45-degree-bank turns again. Slow to 1.3 $V_s$ and try 30-degree-bank turns with half-flaps and a straight-ahead stall with half-flaps. Here again, you are simulating your landing pattern and final approach, but now with landing flaps. Check again for lost altitude. This should help to impress upon your mind the hazards of allowing your aircraft to stall on final approach. Recover from the stall, accelerate to $V_O$, retract your flaps, and continue to accelerate to 100 knots. This will be your maximum speed for the first flight, and you are now ready to go back to the airport. Descend at $V_O$ and 500 feet per minute and enter the pattern. Tell the tower or unicom that you will be making a go-around from short final followed by closed traffic for a full stop. Fly $V_O$ on downwind and lower half-flaps. Slow to $V_O$ minus 5 knots on base to final, and fly this speed over the fence. Execute a go-around, accelerating to $V_O$ and raising the flaps. Fly another pattern to a full-stop landing.

Congratulations! You have just experienced the thrill and satisfaction known only to a small and select group of the world's population. You flew a plane created with your own hands. Now you can gracefully accept the congratulations and praise of your friends while you review the flight in your mind. Then you need to do a thorough postflight and a full structural inspection. Open all access panels and check everything from nose to tail. Carefully correct anything you find wrong and get ready for your next flight.

Subsequent flights will deal with expansion of the aircraft's flight envelope. You will be increasing your speed to 1.1 times the never-exceed speed (1.1 $V_{NE}$), pulling g loads up to your positive g limit, and pushing to your negative g limit. This will let you establish a V-N diagram similar to the example shown below.

As you increase speed to more than 140 knots, you will have to perform flutter tests on your controls all the way up to 1.1 $V_{NE}$ at 10-knot increments. In addition, you will have stability and control testing, stall-spin testing, performance testing, and possibly aerobatic maneuvers to perform. As you can see, whether you are flying a 25-hour or a 40-hour test program, you are going to be busy. You need to plan each flight carefully to make the most of your flight time, and you need to carefully document your tests in your aircraft logbook and in your service and maintenance manual. The details for performing these tests can be found in Vaughan Askue's book, in the *EAA Flight Testing Manual,* or in the FAA Advisory Circular 90-89, Amateur-Built Aircraft Flight Testing Handbook.

When you have completed flight testing, you will be free as a bird (well, almost) to carry passengers and to fly anywhere you wish (within the limits of all applicable FARs) and anytime you wish (if you are equipped for night flying). You still cannot use your homebuilt for any commercial operations, but other than that, you are limited only by the operational capabilities of your aircraft, good judgment, and the FARs. Now go have a ball!

**The Questair Venture looks like a bullet and flies almost as fast.**

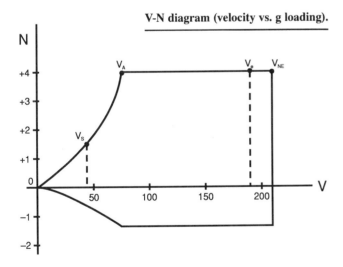

**V-N diagram (velocity vs. g loading).**

| | | |
|---|---|---|
| N | – | load factor (G loading) |
| V | – | velocity (MPH or knots) |
| $V_s$ | – | stall speed at 1G |
| $V_A$ | – | maneuvering speed |
| $V_e$ | – | cruise speed |
| $V_{NE}$ | – | never-exceed speed (red line) |

# 12 The Best Part— Flying It

**N**OW THAT YOU HAVE FLOWN off the test period and investigated the outer limits of your flight envelope, you deserve some time to sit back and enjoy flying your creation. You might want to make the rounds of the summer air shows. You might want to spend a week at Oshkosh or at Sun & Fun in Lakeland, Florida, to show off your superb building skill.

You might want to visit little out-of-the-way airports where airport "bums" will gather around your plane at the gas pumps and ask you a multitude of questions about it. You might want to give rides around the local area to family and friends so they can see what their houses and city look like from the air. Enjoy! You've earned it. But while you're basking in the brilliant glow of your overwhelming accomplishments, don't forget the serious side of flying. Don't forget the professional approach that keeps you from becoming a statistic or the subject of an accident report.

Maintaining your flight proficiency in today's busy world is not always easy. There are many, many things competing for our time, our attention, and our money. You will probably have to make a decidedly conscious effort to give

your flying the time and attention that it deserves. When I get up at 5:00 A.M. to commute to work, I hear a light plane flying just south of our house and heading for the big city. I know someone has found a way to combine business with flying pleasure by commuting to and from work by air while I'm stuck in bumper-to-bumper rush-hour traffic, breathing carbon monoxide fumes for three hours a day. I envy that lucky pilot. If you can find some way to combine your flying with your work, you will be much more likely to remain a highly proficient pilot. If that isn't possible, you will have to maintain your proficiency on your own time.

First and foremost, fly as frequently as you can. And when you do fly, don't just bore holes in the sky. Whenever conditions permit, practice your emergency procedures, run

**79**

A Barracuda in flight over the mountains of southern California. This plane is all-wood construction and capable of limited aerobatics.

How would you like to take this little jewel around the pattern? This is a Fighter Escort Wing (FEW) P-51, a scaled-down replica of the P-51 of World War II fame.

through a stall series, practice engine-out landing approaches, do a few spins if your bird is capable, and always keep an emergency landing spot in the back of your mind. Of course, if you have a passenger on board who turns green in a steep bank, you'll be limited in what you can accomplish, but never let an opportunity to improve your flying skills pass you by. Even on a cross-country flight, you can practice precise flying by holding your altitude, heading, and airspeed as close to the mark as possible while still constantly checking the area for traffic. If you have someone qualified to check the area for you, you can practice instrument flying. How about a little partial panel or needle, ball, and airspeed?

Keep abreast of the latest FAA rule changes and the

Flying this Rotorway Executive "fling wing" is a little different from flying a fixed-wing plane. Rotary-wing birds obviously require special training. You could always join the army to get checked out, but the kit manufacturer can also provide training.

latest happenings in the aviation industry by reading one or two good aviation periodicals every month. Keep your biennial flight review and your medical current. Don't exceed your limitations or the limitations of your aircraft, and remember, the one thing that gets more pilots into trouble than any other single factor is weather, so don't press your luck in adverse weather conditions. Don't run your tanks dry. Don't land with your gear and your head up and locked. The list goes on and on, but it boils down to maintaining a professional attitude toward flying. If you do that, you may still make mistakes, but at least you'll make fewer.

While you are enjoying the miraculous sensations of flight in your very own hand-crafted plane, be sure you are maintaining it properly. Follow a regular preventive maintenance program that keeps everything in good working order between annual inspections. Do a thorough annual on the engine and the airframe. If you are not sure what you should do for an annual inspection, ask an A&P to get you a copy of the checklist used for the annual (or 100-hour inspection) of a factory-built plane similar to your bird, and then modify the list as necessary to fit your inspection. A checklist is an absolute must to ensure that you don't miss anything. Check or initial each item as it is completed and keep the checklist with your other aircraft records at least until the next annual is completed. Make an entry in your logbook for the inspection and for any repairs made. Keep your instruments, radios, and accessory equipment in good condition. Don't fly with inoperable equipment if that equipment is necessary for the safe completion of your flight. Change that ELT (emergency locator transmitter) battery when it is due even if it is still working. Have that propeller overhauled when the TBO (time between overhauls) calls for it. Record all of this good maintenance work that you do in your logbooks. If you add or remove equipment, revise your weight and balance calculations and up-

date the equipment list in your service and maintenance manual.

Many builders never quite finish building their dream machines. Even after they have completed their test period and everything is "in the green," they still want to make their birds better and improve on perfection. They may want to reduce drag, improve engine cooling, or install a more powerful engine. You name it and someone will want to try it. If you are one of these tinkerers, don't forget the cautions discussed previously about modifications to the original design. Be very careful, especially in the areas of aerodynamics, structures, flight controls, and weight and balance. If you are not absolutely certain about the change you are planning, get some expert advice before proceeding.

Then there are those builders who can't wait to get started on the next project. The building becomes the most important thing to them, the proverbial spice of life. The

The Omega II is a real crowd pleaser at any air show, but that 10-gallon hat has got to go before the canopy closes.

Flying ultralights does not require a pilot's license or a medical certificate, but anyone, even a rated pilot, would be well advised to get checked out by a qualified instructor.

flying becomes of secondary or maybe no interest at all. I have known builders who have lost their medicals, but have kept on cranking out homebuilts for the pure joy of seeing an airplane take shape under their capable hands. To them I say, more power to you—keep on building.

For the rest of the throttle benders who can't wait to get in the air, keep on flying. Enjoy the fantastic flying machine you have created to the fullest extent possible. Enjoy it safely and in a professional manner. Take good care of your bird and you'll enjoy it for many, many years. And if the bug bites again, build another, and maybe a third. After all, it's one heck of a lot of fun to build 'em and to fly 'em. So what are you waiting for? Start building! Let's get 'em in the air!

**Here is your cockpit, just waiting for you. All you have to do is build it!**

# List of Sources for Materials, Parts, and Tools

**Aircraft Spruce & Specialty**
P.O. Box 4247
201 W. Truslow
Fullerton, CA 92632
(*material, hardware, kits, books, engines, instruments*)

**Aircraft Steel**
923 W.C.R. #7
Erie, CO 80516
(*sheet metal, tubing*)

**Aircraft Tool & Supply Co.**
P.O. Box 370
2840 Breard St.
Monroe, LA 71211
(*tools*)

**Aircraft Tool & Supply Co.**
P.O. Box 370
1000 Old U.S. 23
Oscods, MI 48750
(*tools*)

**Airparts**
301 N. 7th
Kansas City, KS 66101
(*aluminum sheet metal*)

**Air Tech Coatings**
2300 Redmond Rd.
Jacksonville, AR 72076
(*finishes*)

**Alcor**
P.O. Box 792222
San Antonio, TX 8279-2222
(*instruments*)

**Alexander Aeroplane**
P.O. Box 909
Griffin, GA 30224
(*parts, kits, hardware material*)

**Avery Enterprises**
P.O. Box 387
Bedford, TX 76021
(*tools*)

**B & F Aircraft Supply**
6141 W. 95th St.
Oak Lawn, IL 60453
(*material, tools, accessories*)

**Blue River Aircraft Supply**
P.O. Box 460
Harvard, NE 68944
(*covering supplies*)

**Bon Aero**
7688 Summit Rd.
Parker, CO 80134
(*hardware*)

**BRS, Inc.**
Fleming Field, 1845 Henry Ave.
South St. Paul, MN 55075
(*aircraft recovery systems*)

**California Power Systems**
790 139th, #46
San Leandro, CA 94578
(*engines, accessories, avionics*)

**Century Instrument Corp.**
4440 Southwest Blvd.
Wichita, KS 67210
(*instruments*)

**Chem Tech**
P.O. Box 70148
Seattle, WA 98107
(*T-88 epoxy glue*)

**Chief Aircraft**
Grants Pass Airport
1301 Brookside Blvd.
Grants Pass, OR 97526
(*parts, instruments, accessories*)

**D & D Aircraft Supply**
4 Strickney Terr.
Hampton, NH 03842
(*kitplane and ultralight hardware and fasteners*)

**Decker Aero Power**
Hwy 97 N, Box 495
Marshfield, WI 54449
(*Rotax engines*)

**Electronics International**
5289 NE Elam Young PW, #G200
Hillsboro, OR 97124
(*instruments*)

**Falconer Aviation**
11343 104th St.
Edmonton, Canada T5G 2K7
(*flexible finishes*)

**Full Lotus**
407-5940 #6 Rd.
Richmond, BC, Canada V6V 1Z1
(*floats*)

**Gartmann Stainless**
P.O. Box 1005A
Delran, NJ 08075
(*hardware, fasteners*)

**Gee Bee Canopies**
18415 2nd Ave. S.
Seattle, WA 98148
(*canopies, windows*)

**Harbor Sales Co.**
1401 Russell St.
Baltimore, MD 21230
(*aircraft plywood*)

**J. P. Instruments**
P.O. Box 7033
Huntington Beach, CA 92615
(*instruments*)

**Ken Brock Manufacturing**
11852 Western Ave.
Stanton, CA 90680
(*gyroplane parts, hardware*)

**K. S. Avionics**
25216 Cypress Ave.
Hayward, CA 94544
(*instruments*)

**Leading Edge Airfoils**
331 South 14th St.
Colorado Springs, CO 80904
(*material, hardware, instruments, engines, tools, books, finishes*)

**Neubert Aircraft Supply**
P.O. Box 500
Arroyo Grande, CA 93420
(*hardware, electrical supplies, instruments, avionics*)

**Randolph Products Co.**
Carlstadt, NJ 07072
(*finishes*)

**San-Val Discount**
7444 Valjean Ave.
Van Nuys, CA 91406
(*parts, tools, instruments*)

**Skyline**
McCarran Airport
P.O. Box 98630, 125 E.
Reno Ave.
Las Vegas, NV 89193-8630
(*parts, instruments, avionics*)

**Sky Sports**
Linden-Price Airport, Han
#1
Linden, MI 48451
(*instruments, accessories*)

**Sporty's Pilot Shop**
Dept. SA
Clermont Airport
Batavia, OH 45103
(*instruments, pilot supplies*)

**Stits Aircraft Coatings**
P.O. Box 3084-S
Riverside, CA 92519-3084
(*covering supplies, finishes*)

**Univair Aircraft Corp.**
Rt. 3, Box 59
Aurora, CO 80011
(*hardware, instruments, finishes*)

**USATCO**
Franklin, NY 11010
or Gardena, CA 90948
(*tools, hardware, fasteners*)

**U.S. Industrial Tools**
15119 Cleat St.
Plymouth, MI 48170
(*tools*)

**Van's Aircraft, Inc.**
P.O. Box 160
North Plains, OR 97133
(*kitplanes, parts*)

**Varga Enterprises**
2350 S. Airport Blvd.
Chandler, AZ 85249
(*parts, instruments, avionics, tools, accessories*)

**Wag Aero**
1216 North Rd.
Lyons, WI 53148
(*engines, tools, instruments, avionics, kitplanes*)

**Wicks Aircraft Supply**
410 Pine St.
Highland, IL 62249
(*material, parts, instruments, tools, kitplanes, hardware*)

# APPENDIX B

## U.S. Department of Transportation / Federal Aviation Administration

# ADVISORY CIRCULAR

AC No: 20-27D
Date 6/22/90

## CERTIFICATION AND OPERATION OF AMATEUR-BUILT AIRCRAFT

1. PURPOSE.  This advisory circular (AC) provides guidance concerning the building, certification, and operation of amateur-built aircraft of all types; explains how much fabrication and assembly the builder must do for the aircraft to be eligible for amateur-built certification; and describes the Federal Aviation Administration (FAA) role in the certification process.

2. CANCELLATIONS.  Advisory Circular 20-27C, Certification and Operation of Amateur-Built Aircraft, dated April 1, 1983, is cancelled.

3. BACKGROUND.  The Federal Aviation Regulations (FAR) provide for the issuance of FAA Form 8130-7, Special Airworthiness Certificate, in the experimental category to permit the operation of amateur-built aircraft. Federal Aviation Regulations section 21.191(g) defines an amateur-built aircraft as an aircraft, the major portion of which has been fabricated and assembled by person(s) who undertook the construction project solely for their education or recreation. Commercially produced components and parts which are normally purchased for use in aircraft may be used, including engines and engine accessories, propellers, tires, spring steel landing gear, main and tail rotor blades, rotor hubs, wheel and brake assemblies, forgings, castings, extrusions, and standard aircraft hardware such as pulleys, bellcranks, rod ends, bearings, bolts, rivets, etc.

4. DEFINITION.  As used herein, the term "office" means the FAA Flight Standards District Office (FSDO), Manufacturing Inspection District Office (MIDO), or Manufacturing Inspection Satellite Office (MISO) that may perform the airworthiness inspection and certification of an amateur-built aircraft.

5. FAA INSPECTION CRITERIA.

a. The amateur-built program was designed to permit person(s) to build an aircraft solely for educational or recreational purposes. The FAA has always permitted amateur builders freedom to select their own designs. The FAA does not formally approve these designs since it is not practicable to develop design standards for the multitude of unique design configurations generated by kit manufacturers and amateur builders.

b. In the past, the FAA inspected amateur-built aircraft at several stages during construction. These inspections were commonly called precover inspections. The FAA also inspected the aircraft upon completion, before the initial issuance of the special airworthiness certificate, for the purpose of showing compliance with FAR section 91.42(b) (new FAR section 91.319), and again before issuance of the unlimited duration FAA Form 8130-7. After reassessing the need for these inspections, the FAA in 1983 decided to perform only one inspection prior to the initial flight test.

NOTE: FAR Part 91 has been revised effective August 18, 1990. Both old and new FAR sections are referenced in this AC.

c. Since 1983, FAA inspections of amateur-built aircraft have been limited to ensuring the use of acceptable workmanship methods, techniques, practices, and issuing operating limitations necessary to protect persons and property not involved in this activity.

d. In recent years, amateur builders have adopted a practice whereby they call upon persons having expertise with aircraft construction techniques, such as the Experimental Aircraft Association (EAA) Technical Counselors (reference paragraph 6.(a)) to inspect particular components, e.g., wing assemblies, fuselage, etc., prior to covering, and to conduct other inspections as necessary. This practice is an effective means of ensuring construction integrity.

e. The FAA has designated some private persons to act in its behalf in the inspection of amateur-built aircraft and the issuance

**85**

of airworthiness certificates. These persons are known as Designated Airworthiness Representatives (DAR) and are authorized to charge for their services. These charges are set by the DAR and are not governed by the FAA. The amateur-builder may contact the local FAA office to locate a DAR.

f. In view of the foregoing considerations, the FAA has concluded that safety objectives, relative to the amateur-built program, can be continued to be met by the use of the following criteria:

(1) Amateur builders should have knowledgeable persons (i.e., FAA certified mechanics, EAA Technical Counselors, etc.) perform precover inspections and other inspections as appropriate. In addition, builders should document the construction using photographs taken at appropriate times prior to covering. The photographs should clearly show methods of construction and quality of workmanship. Such photographic records should be included with the builder's log or other construction records.

(2) The FAA inspector or DAR will conduct an inspection of the aircraft prior to the issuance of the initial FAA Form 8130-7 to enable the applicant to show compliance with FAR section 91.42(b) (new FAR section 91.319). This inspection will include a review of the information required by FAR section 21.193, the aircraft builder's logbook, and an examination of the completed aircraft to ensure that proper workmanship has been used in the construction of the aircraft. Also, the appropriate operating limitations will be prescribed at this time in accordance with FAR section 91.42 (new FAR section 91.319).

(3) An FAA inspector or DAR may elect to issue amateur-built airworthiness certificates on a one-time basis to the builder for showing compliance with FAR section 91.42(b) (new FAR section 91.319) and continue operation under FAR section 21.191(g). Under this procedure, the aircraft will be inspected by the FAA only once prior to flight testing. The airworthiness certificate will be issued, but its validity will be subject to compliance with the operating limitations. The limitations will provide for operation in an assigned flight test area for a certain number of hours before the second part of the limitations becomes effective, releasing the aircraft from the test area.

6. <u>DESIGN AND CONSTRUCTION</u>.

a. Many individuals who desire to build their own aircraft have little or no experience with respect to aeronautical practices, workmanship, or design. An excellent source for advise in such matters is the EAA located in Oshkosh, Wisconsin. (See appendix 1.) The EAA is an organization established for the purpose of promoting amateur aircraft building and giving technical advise and assistance to its members. The EAA has implemented a Technical Counselors Program whose aim is to ensure the safety and dependability of amateur-built aircraft. Most EAA Technical Counselors are willing to inspect amateur-built aircraft projects and offer constructive advise regarding workmanship and/or design.

b. Any choice of engines, propellers, wheels, other components, and any choice of materials may be used in the construction of amateur-built aircraft. However, it is strongly recommended that FAA-approved components and established aircraft quality material be used, especially in fabricating parts constituting the primary structure, such as wing spars, critical attachment fittings, and fuselage structural members. Inferior materials, whose identity cannot be established, should not be used. The use of major sections (i.e., wings, fuselage, empennage, etc.) from type certificated aircraft may be used in the construction as long as the sections are in a condition for safe operation. These sections are to be considered by the FAA inspector or DAR in determining the major portion in FAR section 21.191(g), but no credit for fabrication and assembly would be given the builder for these sections. It is recommended that builders contact their local FAA office to coordinate the use of such sections.

c. The design of the cockpit or cabin of the aircraft should avoid, or provide for padding on, sharp corners or edges, protrusions, knobs, and similar objects which may cause injury to the pilot or passengers in the event of an accident. It is strongly recommended that Technical Standard Order (TSO) approved or equivalent seat belts be installed along with approved shoulder harnesses.

d. An engine installation should ensure that adequate fuel is supplied to the engine in all anticipated flight attitudes. Also, a suitable means, consistent with the size and complexity of the aircraft, should be provided to reduce fire hazard wherever possible, including a fireproof firewall between the engine compartment and the cabin. When applicable, a carburetor heat system should also be provided to minimize the possibility of carburetor icing.

e. Additional information and guidance concerning acceptable fabrication and assembly are provided in AC 43.13-1A, Acceptable Methods, Techniques, and Practices - Aircraft Inspection and Repair, and AC 43.13-2A, Acceptable Methods, Techniques, and Practices - Aircraft Alterations. These publications are available from the U.S. Government Printing Office.

f. The builder should obtain the services of a qualified aeronautical engineer or consult with the designer of purchased plans or construction kits, as appropriate, to discuss the proposal if the aircraft design is modified during construction.

7. <u>CONSTRUCTION KITS</u>.

a. Construction kits containing raw materials and some prefabricated components may be used in building an amateur-built aircraft. However, aircraft which are assembled entirely from kits composed of completely finished prefabricated components, parts, and precut and predrilled materials are not considered to be eligible for certification as amateur-built aircraft since the major portion of the aircraft would not have been fabricated and assembled by the builder.

b. An aircraft built from a kit may be eligible for amateur-built certification, provided the major portion has been fabricated and assembled by the amateur builder. Kit owner(s) may jeopardize eligibility for amateur-built certification under FAR section 21.191(g) if they allow someone else to build the aircraft. The major portion of such kits may consist of raw stock such as lengths of wood, tubing, extrusions, etc., which may have been cut to an approximate length. A certain quantity of prefabricated parts such as heat treated ribs, bulkheads or complex parts made from sheet metal, fiber glass, or polystyrene would also be acceptable, provided the kit still meets the major portion of the fabrication and assembly requirement, and the amateur builder satisfies the FAA inspector or DAR that completion of the aircraft kit is not merely an assembly operation.

**CAUTION: Purchasers of partially completed kits should obtain all fabrication and assembly records from the previous owner(s). This may enable the builder who completes the aircraft to be eligible for amateur-built certification.**

c. Various materials/parts kits for the construction of aircraft are available nationally for use by aircraft builders. Advertisements tend to be somewhat vague and may be misleading as to whether a kit is eligible for amateur-built certification. It is not advisable to order a kit before verifying with the local FAA office if the aircraft, upon completion, may be eligible for certification as amateur-built under existing rules and established policy.

d. It should be noted the FAA does not certify aircraft kits or approve kit manufacturers. However, the FAA does perform evaluations of kits which have potential for national sales interest, but only for the purpose of determining if an aircraft built from the kits will meet major portion criteria. A list of these kits is maintained at the local FAA office for information to prospective builders.

8. REGISTRATION AND MARKING INFORMATION. Federal Aviation Regulations section 21.173 requires that all U.S. civil aircraft be registered before an airworthiness certificate can be issued. Federal Aviation Regulations Part 47, Aircraft Registration, prescribes the requirements for registering civil aircraft. The basic procedures are as follows:

a. A builder wishing to register an aircraft must first obtain a registration number assignment (N-number) from the FAA Aircraft Registry. (See appendix 1 for the address of the Aircraft Registry.) This number will eventually be displayed on the aircraft. It is not necessary to obtain a registration number in the early stages of the project. Builders intending to obtain a special number of their choice must submit a letter (see appendix 2) listing up to 5 possible registration numbers desired. Under FAR Part 47, a special registration number will cost $10 and may be reserved for no longer than 1 year. Renewal is necessary each year with an additional $10 fee. If a special number is being requested along with registration, an additional $5 fee is required. Although this reservation does not apply to numbers assigned at random by the Aircraft Registry, it is recommended that application for registration number assignment in either case be made 60 to 90 days prior to completion of construction.

b. To apply for either a random or special registration number assignment, the owner of an amateur-built aircraft must provide information required by the Aircraft Registry by properly completing an Aeronautical Center (AC) Form 8050-88, Affidavit of Ownership for Amateur-Built Aircraft (see appendix 3). The affidavit establishes the ownership of the aircraft; therefore, all aircraft information must be given. If the aircraft is built from an eligible kit, the builder should also submit a signed bill of sale from the manufacturer of the kit as evidence of ownership. If AC Form 8050-2, Aircraft Bill of Sale, is used, the word "aircraft" should be deleted and the word "kit" inserted in its place. (See appendix 4.)

c. After receipt of the applicant's letter requesting a special or random number assignment, the Aircraft Registry will send a form letter to the applicant giving the number assigned (this does not constitute registration of the aircraft), a blank AC Form 8050-1, Aircraft Registration Application, and other registration information. All instructions must be carefully followed to prevent return of the application and delay in the registration process.

d. The applicant must complete and return the white original and green copy of the Aircraft Registration Application (with N-number) to the Aircraft Registry as soon as possible, accompanied with a fee of $5 by check or money order payable to the FAA (see appendix 5). The pink copy of the application is to be retained by the applicant and carried in the aircraft as temporary authority to operate without registration for a maximum of 90 days or until receipt of AC Form 8050-3, Certificate of Aircraft Registration. If AC Form 8050-3 is not received in the 90-day period, the builder must obtain written authority from the Aircraft Registry for continued operation. However, if the recommendations in paragraph 8.a through 8.d above are followed, the applicant should have received AC Form 8050-3 before the airworthiness inspection.

9. IDENTIFICATION AND REGISTRATION MARKS. When applying for an airworthiness certificate for an amateur-built aircraft, the builder must show in accordance with FAR section 21.182 that the aircraft displays the nationality and registration markings required by FAR Part 45. The following is a summary of the FAR Part 45 requirements:

a. The aircraft must be identified by means of a fireproof identification plate that is etched, stamped, engraved, or marked by some approved fireproof marking as required by FAR section 45.11. The identification plate must include the information required by FAR section 45.13.

b. The identification plate must be secured in such a manner that it will not likely be defaced or removed during normal service, or lost or destroyed in an accident. Aircraft built and certificated after March 7, 1988, must have the identification plate located on the exterior either adjacent to and aft of the rear-most entrance door or on the fuselage near the tail surfaces and must be legible to a person standing on the ground (reference FAR section 45.11.)

c. The name on the identification plate must be that of the amateur builder, not the designer, plans producer, or kit manufacturer. The serial number can be whatever the builder wishes to assign, provided it is not the same as other aircraft serial numbers.

d. The builder should refer to FAR sections 45.22 and 45.25, which define specific requirements for the location of registration marks on fixed-wing aircraft. The location of registration marks for non-fixed wing aircraft are specified in FAR section 45.27. These registration marks must be painted on or affixed by means insuring a similar degree of permanence. Decals are also acceptable. The use of tape which can be peeled off, or water soluble paint, such as poster paint, is not considered acceptable.

e. Most amateur-built aircraft are required to display nationality and registration marks with a minimum height of 3 inches.

However, if the maximum cruising speed of the aircraft exceeds 180 knots calibrated air speed (207 miles per hour), the registration marks must be at least 12 inches high. The builder should refer to FAR section 45.29, which defines the minimum size and proportions for nationality and registration marks on all types of aircraft.

f. The registration marks displayed on the aircraft must consist of the Roman capital letter "N" (denoting United States nationality) followed by the registration number of the aircraft. (Registration marks may not exceed five symbols following the prefix letter "N".) These symbols may be all numbers (e.g., N-10000); one to four numbers and one suffix letter (e.g., N-1000A); or one to three numbers and two suffix letters (e.g., N-100AB). Any suffix letter used in the marks must also be a Roman capital letter. The letters "I" and "O" may not be used. The first zero in a number must always be proceeded by at least one of the numbers 1 through 9. In addition, the word "experimental" must also be displayed on the aircraft near each entrance (interior or exterior) to the cabin or cockpit in letters not less than 2 inches nor more than 6 inches in height.

g. If the configuration of the aircraft prevents marking in compliance with any of the above requirements, the builder should contact an FAA office regarding approval of a different marking procedure under FAR section 45.22(d). It is strongly recommended that any questions regarding registration marking be discussed and resolved with a local FAA inspector or DAR before the marks are affixed to the aircraft.

10. CERTIFICATION STEPS. The following procedures are in the general order to be followed in the certification process:

a. Initial Step. The prospective builder should contact the nearest FAA office to discuss the plans for building the aircraft with an FAA inspector. During this contact, the type of aircraft, its complexity and/or materials should be discussed. The FAA may provide the prospective builder with any guidance necessary to ensure a thorough understanding of applicable regulations.

b. Registration. Detailed procedures are in paragraph 8 of this AC. This must be done before submitting an FAA Form 8130-6, Application for Airworthiness Certificate, under FAR section 21.173 to an FAA inspector or DAR.

c. Marking. The registration number (N-number) assigned to the aircraft and an identification plate must be affixed in accordance with FAR section 21.182 and Part 45, Identification and Registration Marking. Detailed procedures are in paragraphs 8 and 9 of this AC.

d. Application. The builder may apply for a special airworthiness certificate by submitting the following documents and data to the nearest FAA office or to the DAR.

(1) Federal Aviation Administration Form 8130-6 (see appendix 6).

(2) Enough data, such as photographs or three-view drawings, to identify the aircraft.

(3) A notarized FAA Form 8130-12, Eligibility Statement - Amateur-Built Aircraft, certifying the major portion was fabricated and assembled for education or recreation, and that evidence is available to support this statement. Evidence will be provided to the FAA Inspector or DAR upon request. (See appendix 7.)

(4) A letter identifying the aircraft and the area over which the aircraft will be tested should accompany the application. (See appendix 8.)

11. AIRCRAFT INSPECTION. The applicant should be prepared to furnish the following to the FAA inspector or DAR:

a. An aircraft complete and ready to fly except for cowlings, fairings, and panels opened for inspection.

b. An Aircraft Registration Certificate, AC Form 8050-3, or the pink copy of Aircraft Registration Application, AC Form 8050-1 (with N-number).

c. Evidence of inspections, such as logbook entries signed by the amateur builder, describing all inspections conducted during construction of the aircraft in addition to photographic documentation of construction details. This will substantiate that the construction has been accomplished in accordance with acceptable workmanship methods, techniques, and practices. It is recommended that this evidence be documented in some form (e.g., the Service and Maintenance Manual available from the EAA).

d. A logbook for the aircraft, engine, and propeller to allow for review of service records and recording of inspection and certification by FAA inspector or DAR.

12. FAA INSPECTION AND ISSUANCE OF AIRWORTHINESS CERTIFICATE

a. After inspection of the documents and data submitted with the application, the applicant should expect the FAA inspector or DAR to inspect the aircraft. Upon determination that the aircraft has been properly constructed, the FAA inspector or DAR may issue an FAA Form 8130-7, together with appropriate operating limitations. The applicant should expect the FAA inspector or DAR to verify that all required markings are properly applied, including the following placard which must be displayed in the cabin or the cockpit at a location in full view of all passengers. (Placard not applicable to single-place aircraft.)

**"PASSENGER WARNING—THIS AIRCRAFT IS AMATEUR-BUILT AND DOES NOT COMPLY WITH FEDERAL SAFETY REGULATIONS FOR STANDARD AIRCRAFT"**

b. Details concerning flight test areas are contained in paragraph 13. The operating limitations are a part of the airworthiness certificate and must be displayed with the certificate when the aircraft is operated. It is the responsibility of the pilot to conduct all flights in accordance with the operating limitations, as well as the General Operating and Flight Rules in FAR Part 91.

c. In the case of a limited duration airworthiness certificate, upon satisfactory completion of operations in accordance with FAR section 91.42(b) (new FAR section 91.319), in the assigned test area, the owner of the aircraft may apply to the local FAA office or DAR for amended operating limitations by submitting another FAA Form 8130-6, along with a letter requesting amendment of operating limitations. Prior to issuance of the amended limitations and a new FAA Form 8130-7, the applicant should expect the

FAA inspector or DAR to review the flight log to determine whether corrective actions have been taken on any problems encountered during the testing and that the aircraft's condition for safe operation has been established. Reinspection of the aircraft may be necessary.

d. Refer to paragraph 13d.(1) and (2) for the processing of unlimited duration airworthiness certificates.

13. FLIGHT TEST AREAS.

a. Amateur-built airplanes and rotorcraft will initially be limited to operation within an assigned flight test area for at least 25 hours when a type certificated (FAA-approved) engine/propeller combination is installed, or 40 hours when a noncertificated (i.e., modified type certificated or automobile) engine/propeller combination is installed. Amateur-built gliders, balloons, dirigibles, and ultralight vehicles built from kits evaluated by the FAR and found eligible to meet requirements of FAR section 21.191(g), for which original airworthiness certification is sought, will be limited to operating within an assigned flight test area for at least 10 hours of satisfactory operation, including at least five takeoffs and landings.

b. The desired flight test area should be requested by the applicant and, if found acceptable by the FAA inspector or DAR, will be approved and specified in the operating limitations. It will usually encompass the area within a 25-statute mile radius (or larger depending on the type of aircraft) from the aircraft's base of operation or in a designated test area established by the local FAA office. The area selected by the applicant and submitted to the FAA for approval should not be over densely populated areas or in congested airways, so that the flight testing during which passengers may not be carried, would not likely impose any hazard to persons on the ground. Advisory Circular 90-89, Amateur-Built Aircraft Flight Testing Handbook, contains recommended procedures for the flight testing of amateur-built aircraft. It is strongly recommended that amateur builders obtain a copy of this AC and follow its guidance.

c. The carrying of passengers will not be permitted while the aircraft is restricted to the flight test area. It is suggested that a tape recorder, for example, be used by the pilot for recording readings, etc. Flight instruction will not be allowed in the aircraft while in the flight test area.

d. In those instances where the unlimited duration special airworthiness certificate was issued, the operating limitations may be prescribed in two phases in the same document as follows:

(1) For phase I limitations, the applicant will receive from the certificating FAA inspector or DAR all those operating limitations, as appropriate, for the applicant to demonstrate compliance with FAR section 91.42(b) (new FAR section 91.319) in the assigned test area. This would further include a limitation requiring the owner/operator to endorse the aircraft logbook with a statement certifying when the aircraft has been shown to comply with FAR section 91.42(b) (new FAR section 91.319). The owner/operator may then operate in accordance with Phase II.

(2) For Phase II limitations, the applicant will receive from the certificating FAA inspector or DAR all those operating limita-

tions, as appropriate, to the issuance of an unlimited duration FAA Form 8130-7 for the operation of an amateur-build aircraft. Appendix 9 contains a sample of typical operating limitations that may be issued. For special conditions, these may vary for each aircraft.

14. SAFETY PRECAUTION RECOMMENDATIONS.
a. All Aircraft.

(1) The pilot should become thoroughly familiar with the brake tests, engine operation, and ground handling characteristics of the aircraft by conducting taxi tests before attempting flight operations. Liftoff is not permitted during taxi tests without an airworthiness certificate.

(2) Before the first flight of an amateur-built aircraft, the pilot should take precautions to ensure that emergency equipment and personnel are readily available in the event of an accident.

(3) Violent or acrobatic maneuvers should not be attempted until sufficient flight experience has been gained to establish that the aircraft is satisfactorily controllable throughout its normal range of speeds and maneuvers. Those maneuvers successfully demonstrated while in the test area may continue to be permitted by the FAA when the operating limitations are modified to eliminate the test area. All maneuvers satisfactorily conducted are to be documented in the aircraft logbook by the pilot.

(4) The operating limitations issued by the FAA inspector or DAR will require the aircraft to be operated in accordance with applicable air traffic control and general operating rules of FAR Part 91 as they apply to amateur-built aircraft. Those operators who plan to operate under Instrument Flight Rules are alerted to the specific requirements under FAR sections 91.115 through 91.129 (new FAR section 91.173 through 91.187).

(5) Depending on the intended operations under FAR Part 91, the following FAR sections may be applicable:

a. FAR section 91.33(b) (new FAR section 91.205) Visual Flight Rules (day).

b. FAR section 91.33(c) (new FAR section 91.205) Visual Flight Rules (night).

c. FAR section 91.33(d) (new FAR section 91.205) Instrument Flight Rules.

(6) Unless authorization to deviate is obtained from Air Traffic Control, any aircraft that will be equipped with a Mode C transponder shall have a calibrated airspeed/static pressure system to prevent an error in altitude reporting. (Reference FAR section 23.1323 and 23.1325.) The Mode C transponder must be tested and inspected per FAR section 91.172 (new FAR section 91.413).

(7) An emergency locator transmitter is required to be on board by FAR section 91.52 (new FAR section 91.207) upon release from the flight test area. Single-place aircraft are exempt from this requirement in accordance with FAR section 91.52(f) (new FAR section 91.207).

b. Rotorcraft. The appropriately rated rotorcraft pilot should be aware of the following operating characteristics:

(1) Operators of rotorcraft having fully articulated rotor systems should be particularly cautious of "ground resonance." This condition of rotor unbalance, if maintained or allowed to progress,

is extremely dangerous and usually results in structural failure.

(2) Tests showing that stability, vibration, and balance are satisfactory should normally be completed with the rotorcraft tied down, before beginning hover or horizontal flight operations.

15. AMATEUR-BUILT AIRCRAFT CONSTRUCTED OUTSIDE THE UNITED STATES AND PURCHASED BY U.S. CITIZENS.

a. When a U.S. citizen purchases such aircraft, acceptable procedures for obtaining airworthiness certification for amateur-built operations are as follows:

(1) The previous owner should have conducted or had a condition/annual type inspection performed on the aircraft within 30 days of the new U.S. owner applying for certification. This inspection shall be recorded in the aircraft records.

(2) The previous owner should obtain documentation from their Civil Aviation Authority that verifies the aircraft is/was originally certificated in that country as an amateur-built, and that the aircraft meets the requirements of FAR section 21.191(g). This documentation should be furnished to the new owner.

b. The new owner of such aircraft shall present the FAA inspector or DAR with a properly completed FAA Form 8130-6, along with the following documentation:

(1) All letters and records of inspections called for in paragraph 15.a. (1) and (2).

(2) Proper documentation of registration in accordance with FAR section 47.

(3) A letter of request to accompany the FAA Form 8130-6.

c. The applicant should expect the FAA inspector or DAR to:

(1) Conduct a thorough review of all documentation called for under paragraphs 15 a and b.

(2) Determine the amateur-built eligibility of the aircraft presented.

(3) Inspect the aircraft like any other amateur-built aircraft, since these airworthiness certifications are considered original.

(4) If the aircraft is found to be eligible and inspection is satisfactory, issue the FAA Form 8130-7 with proper operating limitations. If the required flight time has not been met or there is some question regarding the aircraft's flight capability, the certificating representative may require flight testing.

(5) Advise that the condition inspection on the aircraft can only be performed by the original builder.

16. REPAIRMAN CERTIFICATION. The aircraft builder may be certificated as a repairman if the builder is the primary builder of the aircraft and can satisfactorily prove requisite skill in determining whether the aircraft is in condition for safe operation. This certification can be obtained by making application to the local FAA office after the satisfactory completion of required flight hours in the test area. Each certificate is issued for a particular aircraft. (See appendix 10.)

17. REFERENCE MATERIAL.

a. AC Forms. These forms may be obtained through the local District Office.

AC Form 8050-1, Aircraft Registration Application.

AC Form 8050-2, Aircraft Bill of Sale.

AC Form 8050-88, Affidavit of Ownership for Amateur-Built Aircraft.

FAA Form 8130-6, Application for Airworthiness Certificate.

FAA Form 8130-12, Eligibility Statement - Amateur-Built Aircraft.

FAA Form 8610-2, Airman Certificate and/or Rating Application.

b. Federal Aviation Regulations.

Part 21, Certification Procedures for Products and Parts.

Part 45, Identification and Registration Marking.

Part 47, Aircraft Registration.

Part 65, Certification: Airmen Other Than Flight Crewmembers.

Part 91, General Operating and Flight Rules.

Part 101, Moored Balloons, Kites, Unmanned Rockets, and Unmanned Free Balloons.

Part 103, Ultralight Vehicles.

c. Advisory Circulars.

AC 20-126A, Aircraft Certification Service Field Office Directory.

AC 43.13-1A, Acceptable Methods, Techniques, and Practices - Aircraft Inspection and Repair.

AC 43.13-2A, Acceptable Methods, Techniques, and Practices - Aircraft Alterations.

AC 43-16, General Aviation Airworthiness Alerts.

AC 65-23A, Certification of Repairmen (Experimental Aircraft Builders).

AC 91-23A, Pilot's Weight and Balance Handbook.

AC 183-33A, Designated Airworthiness Representatives.

AC 183-35B, FAA Designated Airworthiness Representatives (DAR), Designated Alteration Stations (DAS), and Delegation Option Authority (DOA) Directory.

18. HOW TO GET PUBLICATIONS. The FAR and those AC's for which a fee is charged may be obtained from the Superintendent of Documents, U.S. Government Printing Office, Washington, DC 20402. A listing of FAR and current prices is in AC 00-44, Status of Federal Aviation Regulations, and a listing of all AC's is in AC 00-2, Advisory Circular Checklist.

19. PUBLICATIONS:

a. To request free advisory circulars, contact:
U.S. Department of Transportation
Utilization and Storage Section, M443.2
Washington, DC 20590

b. To be placed on FAA's mailing list for free AC's contact:
U.S. Department of Transportation
Distribution Requirements
Section, M-494.1
Washington, DC 20590

M.C. Beard
Director, Aircraft Certification Service

**ADDRESSES**

EXPERIMENTAL AIRCRAFT ASSOCIATION, INC. (Telephone, (414) 426-4800); Mail to P.O. Box 3086, Wittman Air Field, Oshkosh, Wisconsin 54903-3086; Street address: 3000 Poberezny Road, Oshkosh, Wisconsin 54903-3086.

FEDERAL AVIATION ADMINISTRATION, AIRCRAFT REGISTRY. (Telephone, (405) 680-3116); Mail to Airman and Aircraft Registry Division, Mike Monroney Aeronautical Center, P.O. Box 25504, Oklahoma City, Oklahoma 73125; Street address: 6500 South MacArthur Boulevard, Oklahoma City, Oklahoma 73169.

**SAMPLE LETTER FOR REQUESTING AN AIRCRAFT REGISTRATION NUMBER
IN ACCORDANCE WITH FAR SECTION 47.15**

<pre>
                                    ___XX-XX-XX___
                                         Date
</pre>

```
FAA Aeronautical Center
FAA Aircraft Registry
P. O. Box 25504
Oklahoma City, Oklahoma  73125

Gentlemen:

This is a request for a United States identification number
assignment for my home-built aircraft.

Aircraft description:

Make __RIGHTWAY__; Type (airplane, rotorcraft, glider, etc.)
__ROTORCRAFT_____; Model ____WHIZ-BANG_____;
Serial number ___001_____.

This aircraft has not been previously registered anywhere.
(FAR section 47.15)


____       Normal Request - $5 (Fee attached)


__X__      Special Registration Number Request -
           $10 (Fee attached)

           CHOICES
           1st __123TR__
           2nd __100TR__
           3rd __100R___
           4th __200TR__
           5th __300TR__




                                    _____
                                         Signature
```

### AFFIDAVIT OF OWNERSHIP FOR AMATEUR-BUILT AIRCRAFT

U.S. Identification Number _____

Builder's Name _JOE BROWN_____

Model _STAR FIRE 1_____ Serial Number _001_____

Class (airplane, rotorcraft, glider, etc.) _AIRPLANE_____

Type of Engine Installed (reciprocating, turbopropeller, etc.)

_____RECIPROCATING_____

Number of Engines Installed _____1_____

Manufacturer, Model and Serial Number of each Engine Installed _____

_____LYCOMING O-2900, 12345_____

_____

Built for Land or Water Operation _____LAND_____

Number of Seats _____2_____

CHECK ONE:

☐ More than 50%) of the above-described aircraft was built from
  ) miscellaneous parts and material kits by the
☐ Less than 50%) undersigned, and I am the owner.


_____
(Signature of Owner-Builder)

State of _____

County of _____

Subscribed and sworn to before me this_____day of_____, 19_____.

My commission expires_____.


_____
(Signature of Notary Public)

AC Form 8050-88 (9-86) (0052-00-559-0002) Supersedes previous edition

## SAMPLE BILL OF SALE

FORM APPROVED
OMB NO. 2120-0042

**UNITED STATES OF AMERICA**
DEPARTMENT OF TRANSPORTATION FEDERAL AVIATION ADMINISTRATION

# KIT ~~AIRCRAFT~~ BILL OF SALE

FOR AND IN CONSIDERATION OF $ THE
UNDERSIGNED OWNER(S) OF THE FULL LEGAL
AND BENEFICIAL TITLE OF THE AIRCRAFT DES-
CRIBED AS FOLLOWS:

UNITED STATES
REGISTRATION NUMBER **N 23456**

AIRCRAFT MANUFACTURER & MODEL
PRATT B-1

AIRCRAFT SERIAL No.
1

DOES THIS          DAY OF          19
HEREBY SELL, GRANT, TRANSFER AND
DELIVER ALL RIGHTS, TITLE, AND INTERESTS
IN AND TO SUCH AIRCRAFT UNTO:

Do Not Write In This Block
FOR FAA USE ONLY

**PURCHASER**

NAME AND ADDRESS
(IF INDIVIDUAL(S), GIVE LAST NAME, FIRST NAME, AND MIDDLE INITIAL.)

BUILDER, EARLY A.
4397 TAKEOFF ROAD
ELROY, AZ 85335

DEALER CERTIFICATE NUMBER

AND TO          EXECUTORS, ADMINISTRATORS, AND ASSIGNS TO HAVE AND TO HOLD
SINGULARLY THE SAID AIRCRAFT FOREVER, AND WARRANTS THE TITLE THEREOF.

IN TESTIMONY WHEREOF          HAVE SET          HAND AND SEAL THIS          DAY OF          19

| | NAME (S) OF SELLER (TYPED OR PRINTED) | SIGNATURE (S) (IN INK) (IF EXECUTED FOR CO-OWNERSHIP, ALL MUST SIGN.) | TITLE (TYPED OR PRINTED) |
|---|---|---|---|
| **SELLER** | HOWARD W. PRATT | | |
| | | | |
| | | | |
| | | | |

ACKNOWLEDGMENT   (NOT REQUIRED FOR PURPOSES OF FAA RECORDING; HOWEVER, MAY BE REQUIRED
BY LOCAL LAW FOR VALIDITY OF THE INSTRUMENT.)

**ORIGINAL: TO FAA**

AC FORM 8050-2 (8-85) (0052-00-629-0002)

## SAMPLE AC FORM 8050-1, AIRCRAFT REGISTRATION APPLICATION

FORM APPROVED
OMB No. 2120-0042

| UNITED STATES OF AMERICA DEPARTMENT OF TRANSPORTATION<br>**FEDERAL AVIATION ADMINISTRATION-MIKE MONRONEY AERONAUTICAL CENTER**<br>AIRCRAFT REGISTRATION APPLICATION | CERT. ISSUE DATE |
|---|---|

| UNITED STATES REGISTRATION NUMBER | N 23456 | |
|---|---|---|
| AIRCRAFT MANUFACTURER & MODEL | PRATT - STEEN | |
| AIRCRAFT SERIAL No. | 001 | FOR FAA USE ONLY |

TYPE OF REGISTRATION (Check one box)

☐ 1. Individual  ☐ 2. Partnership  ☐ 3. Corporation  ☐ 4. Co-owner  ☐ 5. Gov't.  ☐ 8. Non-Citizen Corporation

NAME OF APPLICANT (Person(s) shown on evidence of ownership. If individual, give last name, first name, and middle initial.)

PRATT, ROBERT

TELEPHONE NUMBER: ( 602 ) 412 - 3785

ADDRESS (Permanent mailing address for first applicant listed.)

Number and street: 342 TEABERRY

Rural Route: _____ P.O. Box: _____

| CITY | STATE | ZIP CODE |
|---|---|---|
| SOMEWHERE | ARIZONA | 85000 |

☐ **CHECK HERE IF YOU ARE ONLY REPORTING A CHANGE OF ADDRESS**
**ATTENTION! Read the following statement before signing this application.**
**This portion MUST be completed.**

A false or dishonest answer to any question in this application may be grounds for punishment by fine and / or imprisonment (U.S. Code, Title 18, Sec. 1001).

### CERTIFICATION

I/WE CERTIFY:

(1) That the above aircraft is owned by the undersigned applicant, who is a citizen (including corporations) of the United States.

(For voting trust, give name of trustee: _____ ), or:

CHECK ONE AS APPROPRIATE:

a. ☐ A resident alien, with alien registration (Form 1-151 or Form 1-551) No. _____

b. ☐ A non-citizen corporation organized and doing business under the laws of (state) _____ and said aircraft is based and primarily used in the United States. Records or flight hours are available for inspection at _____

(2) That the aircraft is not registered under the laws of any foreign country; and

(3) That legal evidence of ownership is attached or has been filed with the Federal Aviation Administration.

NOTE: If executed for co-ownership all applicants must sign. Use reverse side if necessary.

TYPE OR PRINT NAME BELOW SIGNATURE

| | SIGNATURE | TITLE | DATE |
|---|---|---|---|
| EACH PART OF THIS APPLICATION MUST BE SIGNED IN INK. | Robert Pratt | OWNER | XX/XX/XX |
| | SIGNATURE | TITLE | DATE |
| | | | |
| | SIGNATURE | TITLE | DATE |
| | | | |

NOTE   Pending receipt of the Certificate of Aircraft Registration, the aircraft may be operated for a period not in excess of 90 days, during which time the PINK copy of this application must be carried in the aircraft.

AC Form 8050-1 (12/90) (0052-00-628-9007) Supersedes Previous Edition

# SAMPLE APPLICATION FOR AIRWORTHINESS CERTIFICATE (AMATEUR-BUILT)

Form Approved
O.M.B. No. 2120-0018

| U.S. Department of Transportation — Federal Aviation Administration | **APPLICATION FOR AIRWORTHINESS CERTIFICATE** | **INSTRUCTIONS** — Print or type. Do not write in shaded areas; these are for FAA use only. Submit original only to an authorized FAA Representative. If additional space is required, use an attachment. For special flight permits complete Sections II and VI or VII as applicable. |
|---|---|---|

## I. AIRCRAFT DESCRIPTION

| 1. REGISTRATION MARK | 2. AIRCRAFT BUILDER'S NAME (Make) | 3. AIRCRAFT MODEL DESIGNATION | 4. YR MFR | FAA CODING |
|---|---|---|---|---|
| N23456 | Pratt | B-1 | 1990 | |

| 5. AIRCRAFT SERIAL NO | 6. ENGINE BUILDER'S NAME (Make) | 7. ENGINE MODEL DESIGNATION | |
|---|---|---|---|
| 1 | Lycoming | 0-290-D | |

| 8. NUMBER OF ENGINES | 9. PROPELLER BUILDER'S NAME (Make) | 10. PROPELLER MODEL DESIGNATION | 11. AIRCRAFT IS (Check if applicable) |
|---|---|---|---|
| 1 | Sensenich | W76AM-2-50 | IMPORT |

## II. CERTIFICATION REQUESTED

**APPLICATION IS HEREBY MADE FOR:** (Check applicable items)

| A | 1 | STANDARD AIRWORTHINESS CERTIFICATE (Indicate category) | NORMAL | UTILITY | ACROBATIC | TRANSPORT | GLIDER | BALLOON |
|---|---|---|---|---|---|---|---|---|

| B | | SPECIAL AIRWORTHINESS CERTIFICATE (Check appropriate items) |

| | 2 | LIMITED | | | |
|---|---|---|---|---|---|
| | 5 | PROVISIONAL (Indicate class) | 1 | CLASS I | |
| | | | 2 | CLASS II | |
| | 3 | RESTRICTED (Indicate operation(s) to be conducted) | 1 | AGRICULTURE AND PEST CONTROL | 2 | AERIAL SURVEYING | 3 | AERIAL ADVERTISING |
| | | | 4 | FOREST (Wildlife conservation) | 5 | PATROLLING | 6 | WEATHER CONTROL |
| | | | 7 | CARRIAGE OF CARGO | 0 | OTHER (Specify) |
| | 4 X | EXPERIMENTAL (Indicate operation(s) to be conducted) | 1 | RESEARCH AND DEVELOPMENT | 2 X | AMATEUR BUILT | 3 | EXHIBITION |
| | | | 4 | RACING | 5 | CREW TRAINING | | MKT SURVEY |
| | | | 0 | TO SHOW COMPLIANCE WITH FAR |
| | 8 | SPECIAL FLIGHT PERMIT (Indicate operation to be conducted, then complete Section VI or VII as applicable on reverse side) | 1 | FERRY FLIGHT FOR REPAIRS, ALTERATIONS, MAINTENANCE OR STORAGE |
| | | | 2 | EVACUATE FROM AREA OF IMPENDING DANGER |
| | | | 3 | OPERATION IN EXCESS OF MAXIMUM CERTIFICATED TAKE-OFF WEIGHT |
| | | | 4 | DELIVERING OR EXPORT | 5 | PRODUCTION FLIGHT TESTING |
| | | | 6 | CUSTOMER DEMONSTRATION FLIGHTS |

| C | 6 | MULTIPLE AIRWORTHINESS CERTIFICATE (Check ABOVE "Restricted Operation" and "Standard" or "Limited" as applicable) |
|---|---|---|

## III. OWNER'S CERTIFICATION

**A. REGISTERED OWNER** (As shown on certificate of aircraft registration) — IF DEALER, CHECK HERE ➤

| NAME | ADDRESS |
|---|---|
| Howard W. Pratt | 1320 West Street, St. Louis MO 41345 |

**B. AIRCRAFT CERTIFICATION BASIS** (Check applicable blocks and complete items as indicated)

| AIRCRAFT SPECIFICATION OR TYPE CERTIFICATE DATA SHEET (Give No and Revision No.) | AIRWORTHINESS DIRECTIVES (Check if all applicable AD's complied with and give latest AD No.) |
|---|---|
| N/A | N/A |

| AIRCRAFT LISTING (Give page number(s)) | SUPPLEMENTAL TYPE CERTIFICATE (List number of each STC incorporated) |
|---|---|
| N/A | N/A |

**C. AIRCRAFT OPERATION AND MAINTENANCE RECORDS**

| CHECK IF RECORDS IN COMPLIANCE WITH FAR 91.173. | TOTAL AIRFRAME HOURS | 3 | EXPERIMENTAL ONLY (Enter hours flown since last certificate issued or renewed) |
|---|---|---|---|
| X | 0.00 | | 0.00 |

**D. CERTIFICATION** — I hereby certify that I am the registered owner (or his agent) of the aircraft described above, that the aircraft is registered with the Federal Aviation Administration in accordance with Section 501 of the Federal Aviation Act of 1958, and applicable Federal Aviation Regulations, and that the aircraft has been inspected and is airworthy and eligible for the airworthiness certificate requested.

| DATE OF APPLICATION | NAME AND TITLE (Print or type) | SIGNATURE |
|---|---|---|
| XX-XX-XX | Howard W. Pratt-owner | Howard W. Pratt |

## IV. INSPECTION AGENCY VERIFICATION

**A. THE AIRCRAFT DESCRIBED ABOVE HAS BEEN INSPECTED AND FOUND AIRWORTHY BY** (Complete this section only if FAR 21.183(d) applies)

| 2 | FAR PART 121 OR 127 CERTIFICATE HOLDER (Give Certificate No.) | 3 | CERTIFICATED MECHANIC (Give Certificate No.) | 6 | CERTIFICATED REPAIR STATION (Give Certificate No.) |
|---|---|---|---|---|---|
| 5 | AIRCRAFT MANUFACTURER (Give name of firm) | | | | |

| DATE | TITLE | SIGNATURE |
|---|---|---|

## V. FAA REPRESENTATIVE CERTIFICATION

(Check ALL applicable blocks in items A and B)

| A. I find that the aircraft described in Section I or VII meets requirements for | | THE CERTIFICATE REQUESTED |
|---|---|---|
| | 4 | AMENDMENT OR MODIFICATION OF CURRENT AIRWORTHINESS CERTIFICATE |
| B. Inspection for a special flight permit under Section VII was conducted by | FAA INSPECTOR | FAA DESIGNEE |
| | CERTIFICATE HOLDER UNDER | FAR 65 | FAR 121, 127 or 135 | FAR 145 |

| DATE | DISTRICT OFFICE | DESIGNEE'S SIGNATURE AND NO | 4 | FAA INSPECTOR'S SIGNATURE | 1 |
|---|---|---|---|---|---|

FAA Form 8130-6 (11-88) SUPERSEDES PREVIOUS EDITION

## SAMPLE ELIGIBILITY STATEMENT AMATEUR-BUILT AIRCRAFT

*Form Approved*
*O.M.B. NO. 2120-0018*

| US Department of Transportation Federal Aviation Administration | **ELIGIBILITY STATEMENT AMATEUR-BUILT AIRCRAFT** | Instructions: Print or type all information except signature. Submit original to an authorized FAA representative. Applicant completes Section I thru III. Notary Public Completes Section IV. |
|---|---|---|

### I. REGISTERED OWNER INFORMATION

Name(s)  E.A. Builder

Address(es)  # 4397 Takeoff Road            Eloy              AZ          85335
                     No. & Street              City        State    Zip

Telephone No.(s) (602) 346-9123                    (602) 346-1253
                  Residence         Business

### II. AIRCRAFT INFORMATION

Model  Vari-Eze                          Engine(s) Make  Lycoming

Assigned Serial No.  EAB-1               Engine(s) Serial No.(s)  803399

Registration No.  W1234B                 Prop./Rotor(s) Make  Sensenich

Aircraft Fabricated:  Plan ☐  Kit ☒      Prop./Rotor(s) Serial No.(s)  479638

### III. MAJOR PORTION ELIGIBILITY STATMENT OF APPLICANT

I certify the aircraft identified in Section II above was fabricated and assembled by                    E.A. Builder
                         Name of Person(s) (Please Print)
for my (their) education or recreation. I (we) have records to support this statement and will make them available to the FAA upon request.

#### — NOTICE —

Whoever in any matter within the jurisdiction of any department or agency of the United States knowingly and willfully falsifies, conceals or covers up by any trick, scheme, or device a material fact, or who makes any false, fictitious or fraudulent statements or representations, or makes or uses any false writing or document knowing the same to contain any false, fictitious or fraudulent statement or entry, shall be fined not more than $10,000 or imprisoned not more than 5 years, or both (U.S. Code, Title 18, Sec. 1001.)

#### APPLICANT'S DECLARATION

I hereby certify that all statements and answers provided by me in this statement form are complete and true to the best of my knowledge, and I agree that they are to be considered part of the basis for issuance of any FAA certificate to me. I have also read and understand the Privacy Act statement that accompanies this form.

Signature of Applicant *(In Ink)*  *Early A. Builder*          Date XX-XX-XX

### IV. NOTARIZATION STATEMENT

FAA Form 8130-12 (4-89)

## SAMPLE LETTER TO ACCOMPANY APPLICATION FOR AIRWORTHINESS CERTIFICATE

To:     (LOCAL FAA OFFICE)          Date: XX-XX-XX
        OR DAR)
        _____
        _____
        _____

In compliance with FAR section 21.193, I hereby request a Special Airworthiness Certificate for my amateur-built aircraft for the purpose of operating amateur-built aircraft.  The aircraft description is as follows:

        Builder: A. BROWN          Registration No:   N6543
          Model: T-BIRD              Serial No:   21
No. of Engines:      1            No. of Seats:   2
Design Criteria; own design _____, plans _____, kit    x

The aircraft has been completely assembled and the following has been accomplished:

Yes  No          I enclose FAA Form 8130-6 which has been completed in Sections I, II, and III.

Yes  No          I enclose FAA Form 8130-12, which has been completed in Sections I, II, and III by me and notarized in Section IV.

Yes  No          I possess AC Form 8050-3 or the pink copy of AC Form 8050-1, signed and dated as evidence that I have complied with the registration requirements of FAR Part 47.

Yes  No          I enclose a three-view drawing or photographs of the aircraft as required by FAR section 21.193.

Yes  No          I have weighed the aircraft to determine that the most forward and aft center of gravity positions are within established limits.  The weight and balance report is available at the aircraft, and a copy is submitted with this application.

Yes  No          I have maintained a construction log for the project, including photographs taken during the construction.  Log entries describe all inspections conducted during construction.

Yes   No       The marking requirements of FAR Part 45 have been
              complied with, including permanent attachment of a
              fireproof aircraft identification (data) plate,
              permanent application of appropriate registration
              marks, and the word "EXPERIMENTAL" near each
              entrance.

Yes   No       The following placard has been displayed in the
              cockpit in full view of all occupants (not
              required for single-place aircraft):

              **"PASSENGER WARNING - THIS AIRCRAFT IS
              AMATEUR BUILT AND DOES NOT COMPLY WITH
              FEDERAL SAFETY REGULATIONS FOR STANDARD
              AIRCRAFT."**

The aircraft will be available for inspection at this location,
and directions are as follows:

GOLD CITY AIRPORT HGR. 5
1400A AIRPORT ROAD
GOLD CITY, NEVADA

I request that the initial operating limitations be issued to
permit me to operate the aircraft within the following
geographical area for flight test:




My residence telephone number is:  __(XXX)  XXX-XXXX__
A daytime telephone number is:      __(XXX)  XXX-XXXX__



_____
Signature (owner/builder)

## SAMPLE LIST OF OPERATING LIMITATIONS

THESE OPERATING LIMITATIONS SHALL BE ACCESSIBLE TO THE PILOT

EXPERIMENTAL OPERATING LIMITATIONS
OPERATING AMATEUR-BUILT AIRCRAFT

REG. NO. _____     SERIAL NO. _____
MAKE     _____     MODEL     _____

Phase I, Initial Flight Test in Restricted Area:

1.  No person may operate this aircraft for other than the purpose of operating amateur-built aircraft to accomplish the operation and flight test outline in the applicant's letter dated _____ in accordance with FAR section 21.193.  Phase I and II amateur-built operations shall be conducted in accordance with applicable air traffic and general operating rules of FAR Part 91 and the additional limitations herein prescribed under the provisions of FAR section 91.42 (new FAR section 91.319).

2.  The initial _____ hours of flight shall be conducted within the geographical area described as follows:

_____

_____

3.  Except for takeoffs and landings, no person may operate this aircraft over densely populated areas or in congested airways.

4.  This aircraft is approved for day VFR operation only.

5.  Unless prohibited by design, acrobatics are permitted in the assigned flight test area.  All acrobatics are to be conducted under the provisions of FAR section 91.71 (new FAR section 91.303).

6.  No person may be carried in this aircraft during flight unless that person is required for the purpose of the flight.

7.  The cognizant FAA office must be notified and their response received in writing prior to flying this aircraft after incorporating a major change as defined by FAR section 21.93.

8.  The operator of this aircraft shall notify the control tower of the experimental nature of this aircraft when operating into or out of airports with operating control towers.

9.  The pilot-in-command of this aircraft must, as applicable, hold an appropriate category/class rating, have an aircraft type rating, have a flight instructor's logbook endorsement or possess a "Letter of Authorization" issued by an FAA Flight Standards Operations Inspector.

10.  This aircraft does not meet the requirements of the applicable, comprehensive, and detailed airworthiness code as provided by Annex 8 to the Convention on International Civil Aviation.  This aircraft may not be operated over any other country without the permission of that country.

Phase II:

Following satisfactory completion of the required number of flight hours in the flight test area, the pilot shall certify in the logbook that the aircraft has been shown to comply with FAR section 91.42(b) (new FAR section 91.319).  Compliance with FAR section 91.42(b) (new FAR section 91.319) shall be recorded in the aircraft logbook with the following or similarly worded statement:

"I certify that the prescribed flight test hours have been completed and the aircraft is controllable throughout its range of speeds and throughout all maneuvers to be executed, has no hazardous operating characteristics or design features, and is safe for operation."

The Following Limitations Apply Outside of Flight Test Area:

1.  Limitations 1, 3, 7, 8, 9, and 10 from Phase I are applicable.

2.  This aircraft is approved for day VFR only, unless equipped for night VFR and/or IFR in accordance with FAR section 91.33 (new FAR section 91.205).

3.  This aircraft shall contain the placards, markings, etc., required by FAR section 91.31 (new FAR section 91.9).

4.  This aircraft is prohibited from acrobatic flight, unless such flights were satisfactorily accomplished and recorded in the aircraft logbook during the flight test period.

5.  No person may operate this aircraft for carrying persons or property for compensation or hire.

6.  The person operating this aircraft shall advise each person carried of the experimental nature of this aircraft.

7.  This aircraft shall not be operated for glider towing or parachute jumping operations, unless so equipped and authorized.

8.  No person shall operate this aircraft unless within the preceding 12 calendar months it has had a condition inspection performed in accordance with FAR Part 43, appendix D, and has been found to be in a condition for safe operation.  In addition, this inspection shall be recorded in accordance with limitation 10 listed below.

9.  The builder of this aircraft, if certificated as a repairman, FAA certified mechanic holding an Airframe and Powerplant rating and/or appropriately rated repair stations may perform condition inspections in accordance with FAR Part 43, appendix D.

10. Condition inspections shall be recorded in the aircraft maintenance records showing the following or a similarly worded statement:

"I certify that this aircraft has been inspected on (insert date) in accordance with the scope and detail of appendix D of Part 43 and found to be in a condition for safe operation."

The entry will include the aircraft total time-in-service, the name, signature, and certificate type and number of the person performing the inspection.


_____          _____
Aviation Safety Inspector                 Date Issued


_____
Office Designation

# SAMPLE—APPLICATION FOR REPAIRMAN AMATEUR-BUILDER

*TYPE OR PRINT ALL ENTRIES IN INK*

Form Approved OMB No. 2120-0022

U.S. Department of Transportation
Federal Aviation Administration

## AIRMAN CERTIFICATE AND/OR RATING APPLICATION

☐ MECHANIC  ☒ REPAIRMAN  ☐ PARACHUTE RIGGER
☐ AIRFRAME  ☐ SENIOR  ☐ MASTER
☐ POWERPLANT  Experimental Aircraft Builder  ☐ SEAT  ☐ CHEST
*(Specify Rating)*  ☐ BACK  ☐ LAP

APPLICATION FOR: ☐ ORIGINAL ISSUANCE  ☐ ADDED RATING

**I. APPLICANT INFORMATION**

**A.** NAME *(First, Middle, Last)*
Charles A. Mayer

**B.** SOCIAL SECURITY NO.  134-90-5210
**C.** DOB *(Mo.,Day Yr.)*  3/15/43
**D.** HEIGHT  70 IN.
**E.** WEIGHT  200

**F.** HAIR  blk
**G.** EYES  hazel
**H.** SEX  M
**I.** NATIONALITY *(Citizenship)*  USA

**J.** PLACE OF BIRTH
Philadelphia, PA

**L.** HAVE YOU EVER HAD AN AIRMAN CERTIFICATE SUSPENDED OR REVOKED?
☒ NO
☐ YES (If "Yes," explain on an attached sheet keying to appropriate item number)

**N.** HAVE YOU EVER BEEN CONVICTED FOR VIOLATION OF ANY FEDERAL OR STATE STATUTES PERTAINING TO NARCOTIC DRUGS, MARIJUANA, AND DEPRESSANT OR STIMULANT DRUGS OR SUBSTANCES? ........ ☒ NO  ☐ YES ⟶

**K.** PERMANENT MAILING ADDRESS
1002 Cable Drive
NUMBER AND STREET, P.O. BOX, ETC.
Oakton
CITY
Virginia  22022
STATE  ZIP CODE

**M.** DO YOU NOW OR HAVE YOU EVER HELD AN FAA AIRMAN CERTIFICATE?  ☐ NO  ☒ YES
SPECIFY TYPE: Private Pilot

DATE OF FINAL CONVICTION

**II. CERTIFICATE OR RATING APPLIED FOR ON BASIS OF —**

☐ **A.** CIVIL EXPERIENCE
☐ **B.** MILITARY EXPERIENCE
☐ **C.** LETTER OF RECOMMENDATION FOR REPAIRMAN *(Attach copy)*

☐ **D.** GRADUATE OF APPROVED COURSE
**(1)** NAME AND LOCATION OF SCHOOL
**(2)** SCHOOL NO
**(3)** CURRICULUM FROM WHICH GRADUATED
**(4)** DATE

☐ **E.** STUDENT HAS MADE SATISFACTORY PROGRESS AND IS RECOMMENDED TO TAKE THE ORAL/PRACTICAL TEST (FAR 65.80)
**(1)** SCHOOL NAME
NO
**(2)** SCHOOL OFFICIAL'S SIGNATURE

☐ **F.** SPECIAL AUTHORIZATION TO TAKE MECHANIC'S ORAL/PRACTICAL TEST (FAR 65.80)
**(1)** DATE AUTH.
**(2)** DATE AUTH. EXPIRES
**(3)** FAA INSPECTOR SIGNATURE
**(4)** FAA DIST. OFC.

**III. RECORD OF EXPERIENCE**

**A.** MILITARY COMPETENCE OBTAINED IN ⟶
**(1)** SERVICE
**(2)** RANK OR PAY LEVEL
**(3)** MILITARY SPECIALTY CODE

**B.** APPLICANTS OTHER THAN FAA CERTIFICATED SCHOOL GRADUATES, LIST EXPERIENCE RELATING TO CERTIFICATE AND RATING APPLIED FOR.
*(Continue on separate sheet, if more space is needed)*

| DATES—MONTH AND YEAR | | EMPLOYER AND LOCATION | TYPE WORK PERFORMED |
|---|---|---|---|
| FROM | TO | | |
| | | | Make—Mayer's Special |
| | | | Model— M-1 |
| | | | Serial No. — 1 |
| | | | Certification, Date of |
| | | | Aircraft— |
| | | | (Date flt. test Complete) XX-XX |

**C.** PARACHUTE RIGGER APPLICANTS: INDICATE BY TYPE HOW MANY PARACHUTES PACKED ⟶
SEAT | CHEST | BACK | LAP

FOR MASTER RATING ONLY
PACKED AS A —
☐ SENIOR RIGGER  ☐ MILITARY RIGGER

**IV. APPLICANT'S CERTIFICATION**
I CERTIFY THAT THE STATEMENTS BY ME ON THIS APPLICATION ARE TRUE
**A.** SIGNATURE
**B.** DATE  XX-XX-XX

**V.** I FIND THIS APPLICANT MEETS THE EXPERIENCE REQUIREMENTS OF FAR 65 AND IS ELIGIBLE TO TAKE THE REQUIRED TESTS.
DATE
INSPECTOR'S SIGNATURE
FAA DISTRICT OFFICE

**FOR FAA USE ONLY**

| Emp. | reg. | D.O. | seal | con | iss | Act | lev | TR | s.h. | Srch | #rte | RATING (1) | RATING (2) | RATING (3) | RATING (4) |
|---|---|---|---|---|---|---|---|---|---|---|---|---|---|---|---|

LIMITATIONS

FAA Form 8610-2 (2-85) SUPERSEDES PREVIOUS EDITION

# APPENDIX C

**U.S. Department of Transportation / Federal Aviation Administration**

# ADVISORY CIRCULAR

AC No: 65-23A
Date 7/22/87

## CERTIFICATION OF REPAIRMEN (EXPERIMENTAL AIRCRAFT BUILDERS)

1. <u>PURPOSE</u>. This advisory circular (AC) provides guidance to builders of experimental aircraft concerning certification as repairmen.

2.. <u>CANCELLATION</u>. AC 65-23, Certification of Repairmen (Experimental Aircraft Builders), dated September 28, 1979, is canceled.

3. <u>RELATED FEDERAL AVIATION REGULATIONS (FAR)</u>. Far Part 21, Sections 21.171, 21.173, 21.175, 21.177, 21.179, 21.181, 21.191, 21.193, and 21.195; FAR Part 43, Appendix D; FAR Part 65, Sections 65.1, 65.11, 65.12, 65.13, 65.15, 65.16, 65.20, 65.21, and 65.104; FAR Part 91, Section 91.42.

4. <u>RELATED MATERIAL</u>. AC 20-27C, Certification and Operation of Amateur-Built Aircraft. Copies of this AC may be obtained by writing to the U.S. Department of Transportation, Utilization and Storage Section, M-443.2, Washington, DC 20590.

5. <u>BACKGROUND</u>.

a. Previously, experimental aircraft certificates were effective for 1 year after the date of issuance or renewal, unless a shorter period was prescribed by the Administrator of the Federal Aviation Administration (FAA). Under the amended provisions of FAR Section 21.181(a)(3), effective September 10, 1979, experimental certificates issued to aircraft for the purpose of exhibition, air racing, or operating amateur-built aircraft have an unlimited duration unless the Administrator finds that a specific period should be established. Thus, performance of recertification inspections on these aircraft by FAA inspectors are no longer required. However, inspectors will continue to perform original certification inspections of experimental aircraft and construction inspections of amateur-built aircraft, amateur-built exhibition aircraft, and air racing aircraft.

b. In conjunction with amended FAR Section 21.181(a)(3), a new FAR Section 65.104, Repairman Certificate (experimental aircraft builder)—Eligibility, Privileges, and Limitations, was added to FAR Part 65. This section provides that a qualified builder of each exhibition, air racing, and amateur-built aircraft may be certificated as a repairman and would be privileged to perform condition inspections in accordance with FAR Part 43, Appendix D. However, aircraft manufacturing companies who produce experimental aircraft are not eligible for repairmen certificates.

c. When provided by the aircraft operating limitations, exhibition, air racing, and amateur-built aircraft may be inspected (condition inspections) by FAA-certificated mechanics holding an airframe and powerplant rating or FAA-certificated and appropriately rated repair stations, in accordance with FAR Part 43, Appendix D.

6. <u>ELIGIBILITY</u>. An individual desiring to be certificated as a repairman is required to:

a. Make application for a repairman certificate (experimental aircraft builder) at the time of original certification of the aircraft. Builders who have had their aircraft certificated prior to the effective date (September 10, 1979) of revised FAR Section 21.181(a)(3) and new FAR Section 65.104 may make application for repairman certification prior to the next condition inspection due date.

b. Be a U.S. citizen or an individual of a foreign country who has been admitted for permanent residence in the United States.

c. Be 18 years of age or older, and the primary builder of the aircraft. For example, when a school, club, or partnership builds an aircraft, only one individual will be considered for a repairman certificate for each aircraft built, such as the class instructor or designated project leader.

d. Demonstrate to the certificating FAA inspector his or her ability to perform condition inspections and to determine whether the subject aircraft is in a condition for safe operation.

Note: The eligibility requirements of FAR Section 65.104 are in no way associated with those eligibility requirements for repairmen shown in FAR Section 65.101, titled "Eligibility Requirements: General."

7. PRIVILEGES AND LIMITATIONS. Holders of repairman certificates (experimental aircraft builder) may perform "condition inspections" on specific aircraft built by the certificate holder. Consistent with FAR 65.104(b), the aircraft will be identified by make, model, and serial number as shown on the repairman certificate. During the aircraft certification procedure, the FAA issues operating limitations, as necessary per FAR Section 91.42, to ensure an adequate level of safety. These limitations may require that the subject aircraft be inspected annually by a repairman (experimental aircraft builder), the holder of an FAA mechanic certificate with appropriate ratings (airframe and powerplant), or an appropriately rated FAA repair station. Condition inspections will be performed in accordance with the scope and detail of FAR Part 43, Appendix D. Operating limitations will also require that an appropriate entry be made in the aircraft maintenance records to show performance of this inspection.

Note: It should be noted that the privileges and limitations of FAR Section 65.104 are not associated with those privileges and limitations of FAR 65.103, titled "Repairman Certificate: Privileges and Limitations."

8. APPLICATION.

a. Applicants may obtain copies of FAA Form 8610-2 (OMB 2120-0022), Airman Certificate and/or Rating Application, from their local FAA General Aviation District Office or Flight Standards District Office. Applicants should complete Items I, III, and IV of this form and submit it to their local FAA office. See Appendix I for an illustrated example. The box for "Repairman" (at top of form) should be checked and underneath in the space for "Specify Rating" print or type the words "Experimental Aircraft Builder." Also, print or type in the "Type of Work Performed" box of Item III the following information relating to the subject amateur-built aircraft:

Aircraft Make_____

Model _____

Serial No. _____

Certification Date of Aircraft _____

Applicants should read the Privacy Act statement attached to FAA Form 8610-2, prior to completing this form.

b. When an applicant meets the certificate eligibility requirements, FAA Form 8060-4, Temporary Airman Certificates, will be issued. Permanent certificates will be mailed to the holder of a Temporary Airman Certificate within 120 days of issuance.

9. SURRENDERED CERTIFICATE PROCEDURES. Repairman certificates (experimental aircraft builder) should be surrendered whenever the aircraft is destroyed or sold. However, in the latter situation, the repairman may elect to retain the certificate in order to perform condition inspections on the aircraft for the new owner. Surrendered certificates should be forwarded to the Mike Monroney Aeronautical Center, Airmen Certification Branch, AAC-260, P.O. Box 25082, Oklahoma City, Oklahoma 73125, with a brief statement of reasons for surrender.

10. TYPICAL AIRCRAFT OPERATING LIMITATIONS. The following or similarly worded aircraft operating limitations may be issued at the time of aircraft certification:

a. No person may operate this aircraft unless within the preceding 12 calendar months it has had a condition inspection performed in accordance with FAR Part 43, Appendix D, and is found to be in a condition for safe operation. Additionally, this inspection shall be recorded in accordance with the limitation in subparagraph d.

b. For amateur-built aircraft, amateur-built exhibition aircraft, and air racing aircraft: Only FAA-certificated repairmen (show repairman's name), mechanics holding an airframe and powerplant rating, and appropriately rated repair stations may perform condition inspections in accordance with FAR Part 43, Appendix D.

c. For other exhibition and air racing aircraft: Only FAA-certificated and rated airframe and powerplant mechanics and appropriately rated repair stations may perform condition inspections in accordance with FAR Part 43, Appendix D.

d. Condition inspections shall be recorded in the aircraft maintenance records showing the following or a similarly worded statement: "I certify that this aircraft has been inspected on (insert date) in accordance with the scope and detail of FAR Part 43, Appendix D, and found to be in a condition for safe operation." The entry will include the aircraft total time-in-service, the name, signature, and certificate type and number of the person performing the inspection.

William T. Brennan
Acting Director of Flight Standards

## SAMPLE FAA FORM 8610-2, AIRMAN CERTIFICATE AND/OR RATING APPLICATION (FRONT)

**TYPE OR PRINT ALL ENTRIES IN INK**

Form Approved OMB No. 2120-0022

U.S. Department of Transportation
Federal Aviation Administration

### AIRMAN CERTIFICATE AND/OR RATING APPLICATION

- [ ] MECHANIC
  - [ ] AIRFRAME
  - [ ] POWERPLANT
- [X] REPAIRMAN

  Experimental Aircraft Builder
  *(Specify Rating)*
- [ ] PARACHUTE RIGGER
  - [ ] SENIOR
  - [ ] SEAT
  - [ ] BACK
  - [ ] MASTER
  - [ ] CHEST
  - [ ] LAP

APPLICATION FOR: [ ] ORIGINAL ISSUANCE   [ ] ADDED RATING

### I. APPLICANT INFORMATION

| A. NAME *(First, Middle, Last)* | | | | K. PERMANENT MAILING ADDRESS |
|---|---|---|---|---|
| Charles Mayer | | | | 1002 Cable Drive |

| B. SOCIAL SECURITY NO. | C. DOB *(Mo.,Day Yr.)* | D. HEIGHT | E. WEIGHT |
|---|---|---|---|
| 13490521 | 3/15/43 | 70 IN. | 200 |

NUMBER AND STREET, P.O. BOX, ETC.

Oakton, Virginia 22022

| F. HAIR | G. EYES | H. SEX | I. NATIONALITY *(Citizenship)* |
|---|---|---|---|
| Black | Hazel | M | USA |

CITY

Fairfax

J. PLACE OF BIRTH

Philadelphis, PA

STATE                ZIP CODE

L. HAVE YOU EVER HAD AN AIRMAN CERTIFICATE SUSPENDED OR REVOKED?
[X] NO
[ ] YES *(If "Yes," explain on an attached sheet keying to appropriate item number)*

M. DO YOU NOW OR HAVE YOU EVER HELD AN FAA AIRMAN CERTIFICATE?   [X] NO   [ ] YES
SPECIFY TYPE:

N. HAVE YOU EVER BEEN CONVICTED FOR VIOLATION OF ANY FEDERAL OR STATE STATUTES PERTAINING TO NARCOTIC DRUGS, MARIJUANA, AND DEPRESSANT OR STIMULANT DRUGS OR SUBSTANCES? . . . . . . . . . . . . . . . . . . . . . . . . . . . . . . . . . [X] NO   [ ] YES ——→

DATE OF FINAL CONVICTION

### II. CERTIFICATE OR RATING APPLIED FOR ON BASIS OF —

- [ ] A. CIVIL EXPERIENCE
- [ ] B. MILITARY EXPERIENCE
- [ ] C. LETTER OF RECOMMENDATION FOR REPAIRMAN *((Attach copy)*

| [ ] D. GRADUATE OF APPROVED COURSE | (1) NAME AND LOCATION OF SCHOOL | | |
|---|---|---|---|
| | (2) SCHOOL NO. | (3) CURRICULUM FROM WHICH GRADUATED | (4) DATE |

[ ] E. STUDENT HAS MADE SATISFACTORY PROGRESS AND IS RECOMMENDED TO TAKE THE ORAL/PRACTICAL TEST (FAR 65.80)

| (1) SCHOOL NAME | NO | (2) SCHOOL OFFICIAL'S SIGNATURE |
|---|---|---|

[ ] F. SPECIAL AUTHORIZATION TO TAKE MECHANIC'S ORAL/PRACTICAL TEST (FAR 65.80)

| (1) DATE AUTH. | (2) DATE AUTH. EXPIRES | (3) FAA INSPECTOR SIGNATURE | (4) FAA DIST. OFC. |
|---|---|---|---|

### III. RECORD OF EXPERIENCE

A. MILITARY COMPETENCE OBTAINED IN ——→

| (1) SERVICE | (2) RANK OR PAY LEVEL | (3) MILITARY SPECIALTY CODE |
|---|---|---|

B. APPLICANTS OTHER THAN FAA CERTIFICATED SCHOOL GRADUATES, LIST EXPERIENCE RELATING TO CERTIFICATE AND RATING APPLIED FOR.
*(Continue on separate sheet, if more space is needed)*

| DATES—MONTH AND YEAR | | EMPLOYER AND LOCATION | TYPE WORK PERFORMED |
|---|---|---|---|
| FROM | TO | | |
| | | | Make – Mayer's Special |
| | | | Model – M-1 |
| | | | Serial No. – No. 1 |
| | | | Certification Date of Aircraft |
| | | | May 2, 1992 |

| C. PARACHUTE RIGGER APPLICANTS: INDICATE BY TYPE HOW MANY PARACHUTES PACKED ——→ | SEAT | CHEST | BACK | LAP | FOR MASTER RATING ONLY | PACKED AS A — |
|---|---|---|---|---|---|---|
| | | | | | | [ ] SENIOR RIGGER   [ ] MILITARY RIGGER |

### IV. APPLICANT'S CERTIFICATION

I CERTIFY THAT THE STATEMENTS BY ME ON THIS APPLICATION ARE TRUE

| A. SIGNATURE | B. DATE |
|---|---|
| *Charles Mayer* | December 30, 1992 |

### V.

I FIND THIS APPLICANT MEETS THE EXPERIENCE REQUIREMENTS OF FAR 65 AND IS ELIGIBLE TO TAKE THE REQUIRED TESTS.

| DATE | INSPECTOR'S SIGNATURE | FAA DISTRICT OFFICE |
|---|---|---|

**FOR FAA USE ONLY**

| Emp. | reg. | D.O. | seal | con | iss | Act | lev | TR | s.h. | Srch | #rte | RATING (1) | RATING (2) | RATING (3) | RATING (4) |
|---|---|---|---|---|---|---|---|---|---|---|---|---|---|---|---|

LIMITATIONS

FAA Form 8610-2 (2-85) SUPERSEDES PREVIOUS EDITION

# APPENDIX D

## CANADIAN RULES FOR HOMEBUILTS

### THE EXPERIMENTAL AIRCRAFT ASSOCIATION CANADIAN COUNCIL (EAACC)

The Experimental Aircraft Association Canadian Council was originally formed in the early 60s as an umbrella organization for members of the Experimental Aircraft Association and the EAA Chapters in Canada. In 1967, the EAA Canadian Council was incorporated as EAA Canada (EAAC). In February 1987, a special EAAC Directors' Meeting discontinued the amalgamation of EAAC with all groups and umbrella organizations and in the summer of 1988 changed the name of the organization that had been known as EAA Canada to the Recreational Aircraft Association of Canada (RAAC). Following the dis-affiliation from the Experimental Aircraft Association by the EAAC/RAAC, the EAA Canadian Council was re-formed for those who wished to continue a close working relationship with the Experimental Aircraft Association. The EAA Canadian Council (EAACC) operates as a committee of Canadian EAA members funded by EAA International in Oshkosh and encourages participation in the programs and activities offered by EAA with emphasis on youth, education, and safety in being of assistance to Canada's aviation community.

For a complete list of the services provided by EAA; magazines, educational programs, merchandise, videos, photo reprints, auto fuel STCs, EAA divisions, EAA Chapter programs, library services, Air Adventure Museum, etc., please write the EAA Canadian Council and ask for the *Membership Information Booklet*.

The EAA Canadian Council can provide you with names and addresses of registered owners of amateur-built aircraft. You may then contact them to find out what they think about the aircraft you may be thinking of building. A maximum of three designs for each request, please.

### CANADIAN OWNERS AND PILOTS ASSOCIATION (COPA)

While EAA and the EAA Canadian Council can be of assistance and encouragement to you in the building of your amateur-built aircraft, when the wheels leave the ground, you and your aircraft are subject to the same rules and regulations as an aircraft with an ATA (Aircraft Type Approval). For this reason, the EAA Canadian Council encourages you to support the Canadian Owners and Pilots Association, an organization with over forty years of experience in crusading for your rights as a pilot and aircraft owner.

COPA is Canada's umbrella organization with representation on the COPA Board of Directors from the 99s, Seaplane Pilots Association, Flying Farmers, RAAC, and the EAA Canadian Council. The representatives are appointed by the organization in question and have full voting privileges on items before the board.

COPA publishes a monthly section in its yellow pages of interest to homebuilders, *The Canadian Homebuilt Aircraft News*. If you would like to receive a complimentary copy, please write to:

**The EAA Canadian Council**
Box 734, Stn B
Ottawa, Ontario K1P 5S4

### CANADIAN ORGANIZATIONS, PUBLICATIONS, SUPPLIERS, AND SOURCES OF MATERIALS

**Western Aircraft Supplies**
623 Markerville Rd. N.E.
Calgary, Alta. T2E 5X1
403-276-3087

**Falconar Aviation Ltd.**
19 Airport Rd.
Edmonton, Alta. T5G 0W7
403-454-7272

**Leavens Aviation Ltd.**
2555 Derry Rd.East
Mississauga, Ontario L4T 1A1
416-678-1234

**Leavens Aviation Ltd.**
Han 3 Industrial AP
Edmonton, Alta. T2G 2Z3
403-477-5568

**Salvair Limited**
R.R. 1
Beaverton, Ontario L0K 1A0
705-426-5135

**Barn Full of Parts**
R.R. 2
Hamilton, Ontario L8N 2Z7
416-689-8257

**Grass Roots Aviation**
648 Adelaide Avenue West
Oshawa, Ontario L1J 6S2
416-434-4651

**Superflite Canada**
2857 Derry Rd. Suite 450
Mississauga, Ont. L4T 1A6
416-677-4112

**Canadian Owners & Pilots Asso.**
Box 734, Stn B
Ottawa, Ontario K1P 5S4
613-236-4901

**Ultra-Light Pilots Asso.**
Box 116
Baldwin, Ontario L0E 1A0
416-722-7267

**109**

**Canadian Seaplane Pilots Asso.**
201 Consumers Rd. Suite 105
Willowdale, Ontario M2J 4G8
416-498-1555

**International Flying Farmers**
Box 9124, 2120 Airport Rd.
Mid-Continent Airport
Wichita, KS 67277
316-943-4234
(Chapters in Canada)

**Ninety-Nines**
The East Canada Section Inc.
c/o The Prop Shop
Buttonville Airport
Markham, Ont. L3P 3J9
416-889-6788

Before starting construction of your amateur-built aircraft, write to the nearest Transport Canada Aviation Group (TCAG) office to obtain an application form. Complete the form and return it to TCAG. TCAG will give you details on the inspections that must be carried out during construction. Inspections will be done either by a TCAG Airworthiness Inspector or by a Delegated Amateur-Built Inspector (DABI) depending on where you live in Canada.

What is an amateur-built aircraft? You or a group of individuals build an aircraft from raw materials on a non-commercial, non-production basis for educational or recreational purposes only. You must build at least 51% of the aircraft from raw materials and the other 49% can be built by the manufacturer of a kit. Methods of fabrication, assembly, and workmanship must meet accepted aviation standard practices for materials used, hardware specifications, etc. You can use any type of powerplant using a propeller or reaction jet propulsion, but you cannot use solid and liquid fuel rockets. There are requirements for weight, maximum wing loadings, minimum rated engine power, etc. These are detailed in the following publication that you may purchase for your own use. Chapter 549, Amateur-Built Aircraft Airworthiness Standards Manual, Catalog T51-13/549E, may be ordered from:

**Canadian Government Publishing**
Ottawa, Ontario
K1A 0S9

Send a cheque or money order to the Receiver General for $17.39 to cover the cost of the manual, tax, and postage.

The following is a summary of the requirements for a fixed wing, standard performance aircraft as specified in Chapter 549:

Maximum of four seats
Maximum gross weight: 3,968 pounds (1,800 kg)
Maximum wing loading, no flaps: 13.3 lb/sq ft (65 kg/sq m)
Maximum wing loading, flaps: 20.4 lb/sq ft (100 kg/sq m)
Safety belts required
Firewall standards apply
Identification plate required
Logs and maintenance records required
Pilot license required
Inspections required during building
Various placards needed
Carburetor ice prevention required
Initial operating restrictions apply

Various requirements for instruments, equipment, design changes, weight and balance, etc.

Amateur-built aircraft that exceed the wing loadings noted above are classed as high performance aircraft and require a cockpit placard as noted in Chapter 549 and a high performance type rating on the license of any pilot flying the aircraft.

The following forms are required for amateur-built aircraft:

Application to Construct an Amateur-Built Aircraft
Application for Special Certificate of Airworthiness
Weight and Balance Report
Climb Test Report
Application for Registration

These forms are available from any TCAG office. A free booklet, *Civil Aviation Publications TP3680E* is available from:

**Transport Canada AANDHD**
Ottawa, Ontario
K1A 0N8

The following requirements apply to ultra-light aircraft:
Maximum launch weight, single place: 363 pounds
Maximum launch weight, two place: 430 pounds
Maximum wing loading: 5.14 lb/sq ft
Minimum wing area: 107.6 sq ft
A fee is required for registration marks, C-Ixxx
Identification plate required
No Certificate of Airworthiness or Flight Permit required
Operate in day VFR only
No logs required
No ELT required
Ultra-light pilot's license is required
Two place for dual instruction only

There are conditions regarding flight into controlled airspace, etc. For complete details, obtain the *Ultra-Light Aeroplane and Hang Glider Information Manual TP4310E* for $5.35 (includes GST) from Transport Canada (address above). The manual includes information on regulations, exemptions, licensing, registration and marking, Canadian airspace, aircraft radios, airports, etc.

As of January 1, 1993, the following requirements went into affect for advanced ultra-light aeroplanes (AULA):

Maximum single place gross weight: 628.3 pounds
Maximum two place gross weight: 1058.2 pounds
Maximum stall speed at gross weight: 45 MPH
Registration marks: C-Fxxx or C-Gxxx
Category 4 medical required for private pilot
Category 3 medical required for commercial pilot

AULA may be flown in controlled airspace if they meet the required equipment orders and are flown by a suitably rated pilot.

Commercial pilot's license required to operate an AULA with a passenger or in controlled airspace.

Private pilot's license required to operate an AULA without passengers and in uncontrolled airspace outside a five nautical mile radius from an airport unless prior permission is received.

There are three ways to obtain an AULA. It may be purchased from a factory already assembled and test flown. In this case, the AULA will operate under a Certificate of Registration which is issued on the basis of meeting the design standards of TP10141 as determined by a Statement of Conformity. If the Statement of Conformity becomes invalid for any reason, the Certificate of Registration is deemed to be cancelled. The AULA purchased from a factory may be used for flight instruction and may be rented to qualified pilots.

If an AULA is built from a factory kit, initial operating restrictions apply for the first five hours of flight and no passengers can be carried during this period. Also, no flight over built up areas or open air assemblies of people are allowed during the initial five hours of flight.

If an AULA is built in accordance with Chapter 549, at least 51% must be constructed by the builder and initial operating restrictions apply for the first five hours of flight as above. In this case, the AULA will operate under a Special Certificate of Airworthiness and allowable operations will be determined by the rating of the pilot flying the AULA.

A complete copy of the AULA policy is provided as a service by the EAA Canadian Council at a cost of $7.00 to cover the cost of copying, taxes, binding, and postage.

Can aerobatics be performed in amateur-built aircraft in Canada? Some amateur-built aircraft have been evaluated and found suitable for aerobatic flight. These are: the Steen Skybolt, Zenair Zenith (with aerobatic spar), Wag-Aero Acro Trainer, Jenkinson FJ Special, Zenair Acro Zenith, Super Acro Zenith, Ultimate Pitts, some models of the Pitts Special, Christen Eagle, and Acro Avia (Les Mitchell's). One of the funtions of the Canadian Aerosport Technical Committee (CASTC) is the evaluation of amateur-built aircraft for aerobatics. This volunteer group of professionals will provide precise details on th capabilities and the shortcomings of a design. Please note that the removal of aerobatic restrictions, as noted in Chapter 549, does not guarantee aerobatic capability, but is to be taken as an indication of structural and aerodynamic capability to survive the aerobatic environment. Each design is evaluated on its own merit. The structural evaluation covers limit load factor classifications: sportsman, intermediate, advanced, and unlimited, along with the documentation and structural analysis. A mechanical evaluation of cockpit controls, systems, and weight and balance is part of the package. The flight evaluation covers the flight envelope, stability and control, flutter, vibration and buffeting, stall and departure maneuvers, stalls, spins, and various aerobatic maneuvers.

On receipt of the report by the Canadian Aerosport Technical Committee, Transport Canada will determine if the aircraft is suitable for aerobatics in Canada as an amaetur-built aircraft. If it is, the restriction, "Aerobatics Prohibited", will be removed from the Special Certificate of Airworthiness for the amateur-built aircraft.

If you are looking for a reference manual that will give you over 600 pages of information on almost 1600 amateur-built aircraft types by nearly 700 designers, the Canadian Aerosport Technical Committee has the answer. The current edition of the *Amateur-Built Aircraft Reference Manual CASTC TIR-6*, lists all current amateur-built aircraft types or models, sources of plans and kits, type newsletters, and other useful information. It can also be used as a source catalog. The manual lists many one-off types for which plans and/or kits are no longer available. The manual lists almost 2600 aircraft, 830 of which are no longer flying in Canada and have been dropped from Transport Canada's Canadian Civil Aircraft Register. There are over 5200 references to other publications where additional information is available. If you attended EAA Oshkosh '91, you likely had the opportunity to see the copy on hand at the Homebuilder's Corner. Ted Slack is also taking a copy with him as he travels in Canada. The reference manual is the result of many years of dedicated effort by Bill Laundry. The manual, in three volumes, is available for a cost of $60 Canadian or $55 U.S. Prices include shipping, handling, and taxes. Send a check or a money order to:

**Canadian Aerosport Technical Committee**
Suite 201
15 Grenfell Crescent
Nepean, Ontario K2G 0G3

*Airworthiness Standards: Amateur-Built Aircraft,* Chapter 549, details the log books that are required for your amateur-built aircraft—a Journey Log Book and Tecnical Log Books. In order to show the inspector that you have built more than 50% of the aircraft, it is strongly recommended that you keep a construction log with daily entries of construction and photographs taken as the major components are completed. Bills of sale for material and parts may be maintained in a three-ring binder with divisions for the aircraft, engine, and propeller. With this information, you can answer any questions by the inspector as to the sources of material and specifications of parts. For example, life-limited parts such as helicopter blades must have a paper trail.

The Journey Log must be kept on board your aircraft. It records the flight history and documents the nature and duration of each flight. If your aircraft has had the aerobatic limitations removed, the log should record the demonstrated aerobatic maneuvers carried out. The reading of the G meter can be recorded as well. The Journey Log will substantiate the completion of the minimum period of flight time required to rectify design and/or construction errors plus the additional twenty-five hours of trouble free operation. This varies. Fixed wing aircraft are twenty-five hours; fixed wing gliders are ten hours.

Technical Logs are required in accordance with ANOs and shall contain particulars of any repair to, modification of, defect of, airframe, engine, propeller, or any component, as applicable for the type of aircraft. If you sell the aircraft, the logs shall be passed to the new owner(s) of the aircraft.

Federal and Provincial governments have a working arrangement whereby newly registered aircraft (including amateur-builts) are checked to see that applicable taxes have been paid. If taxes are not shown to have been paid for the registration mark for your newly completed amateur-built, you will receive a request to pay the taxes or to show from your receipts, that the taxes were paid at the time of purchase for the material and components in your aircraft. In some cases, the Provincial tax department will ask you to pay the taxes on items you have purchased out of Province and/or on which Provincial taxes have not been paid. Requirements vary from Province to Province. If you cannot produce bills of sale for the materials and components in your aircraft and cannot otherwise show the manufactured cost of the aircraft you have built, some tax departments may arbitrarily place a value on the aircraft and taxes will apply on that value. You can save yourself a lot of hassle (and perhaps money) by keeping good records of what you build and what you spend. Unfortunately, as one builder said, "When the weight of the paperwork equals the weight of the aircraft, it is ready to fly!"

The regular inspection of your amateur-built aircraft or your amateur-built advanced ultra-light aeroplane will be carried out either by a Transport Canada Aviation Group (TCAG) Airworthiness Inspector or by a Delegated Amateur-Built Inspector (DABI). The waiting time for a TCAG Inspector is as noted below. There is no cost to you for inspections by a TCAG Inspector, but there could be a waiting time as noted below for the various regions in Canada. Inspections by a DABI are conducted by the Recreational Aircraft Association of Canada (formerly EAAC), 152 Harwood Avenue South, Ajax, Ontario L1S 2H6. There is a fee for each inspection ranging from $125 to $200. Travel expenses for the DABI to do the inspection may also be a consideration. You are not required to be a member of the RAAC, or any organization, in order to have your aircraft inspected. The inspection program is operated by the RAAC on a "fee for services" basis. At present, it is operating in most parts of Ontario.

Before starting construction of your amateur-built aircraft or your amateur-built advanced ultra-light aeroplane, contact your nearest Transport Canada regional office as noted below, to obtain permission to build your aircraft and to receive information as to who will be carrying out the inspections on your aircraft.

Pacific Region
Transport Canada Aviation
Suite 620
800 Burrard Street
Vancouver, British Columbia
V6Z 2J8

Because of the many high priority demands on our resources, amateur-built inspections are undertaken on an opportunity basis only. However, under local arrangements we accept recommendations from retired inspectors who are prepared to undertake the work on behalf of an applicant. This has provided the amateur builders in this region with a means of obtaining prompt inspection service.

Central Region
Transport Canada Aviation
333 Main Street
P.O. Box 8550

Winnipeg, Manitoba
R3C 0P6

For local amateur-built aircraft up to five days wait for inspection. For other areas, inspection time would vary from one to six weeks depending on the location and programming with other work in that area, and the amount of advance notice received.

Quebec Region
Transport Canada Aviation
Regional Administration Building
Montreal International Airport
P.O. Box 500
Dorval, Quebec
H9R 5I8

Since a planning of inspections has been established to a level of four units per month for homebuilt aircraft, a service delay of 90 days can be envisaged.

Western Region
Transport Canada Aviation
Canada Place
1100-9700 Jasper Avenue
Edmonton, Alberta
T5J 4E6

These inspections vary more widely than is the case for other aircraft. If an amateur-built aircraft is being constructed in an area where there is very little commercial aviation activity, the wait can be between two weeks to six months, or even more in some cases. Western Region covers a very large geographical area and we might not always be able to respond as quickly as would be possible in a smaller, frequently travelled region.

Ontario Region
Transport Canada Aviation
4900 Yonge Street
Suite 300
Willowdale, Ontario
M2N 6A5

Homebuilt inspections: Since the inception of the Designated Amateur-Built Inspection Program in the Ontario Region some ten months ago, very few requests have been received by this office for this service. Any requests received, however, would receive very low priority and would not be completed until the next scheduled visit to the location of the aircraft.

Atlantic Region
Transport Canada Aviation
P.O. Box 42
Moncton, New Brunswick
E1C 8K6

The average time to respond to requests for amateur-built aircraft inspections is three to four weeks, depending on our workload.

# BIBLIOGRAPHY

Askue, Vaughan. *Flight Testing Homebuilt Aircraft*. Ames, IA, 1992.

Bingelis, Tony. *The Sportplane Builder*. Austin, TX, 1979.

———. *Firewall Forward*. Austin, TX, 1983.

———. "About That Instrument Panel. . . ." *Sport Aviation* 41, no 11 (1992): 85-90.

Bowers, Peter M. *The 25 Most Practical Homebuilt Aircraft*. Blue Ridge Summit, PA, 1978.

———. *Guide to Homebuilts*. Blue Ridge Summit, PA, 1981.

Caidin, Martin Strasser. *Test Pilots: Riding the Dragon*. New York, 1992.

Davisson, Budd. *The World of Sport Aviation*. New York, 1982.

Downie, Don and Julia. *Complete Guide to Rutan Homebuilt Aircraft*. Blue Ridge Summit, PA, 1981.

Markowski, Michael. *The Encyclopedia of Homebuilt Aircraft*. Blue Ridge Summit, PA, 1980.

Matricardi, Paolo. *The Concise History of Aviation*. New York, 1984.

Nelson, John L. *Lightplane Engines*. Blue Ridge Summit, PA, 1981.

Sport Pilot. "200 Sport Planes for Under $15,000." *Sport Pilot* 3 (1988): 2-90.

U.S. Department of the Air Force. *Instrument Flying*. Washington, DC, 1960.

*World Book Encyclopedia*. Chicago, 1991 ed.

# INDEX

**115**